OZONE DEPLETION, GREENHOUSE GASES, AND CLIMATE CHANGE

Proceedings of a Joint Symposium by the
Board on Atmospheric Sciences and Climate
and the
Committee on Global Change
Commission on Physical Sciences, Mathematics, and Resources
National Research Council

NATIONAL ACADEMY PRESS
Washington, D.C. 1989

National Academy Press • 2101 Constitution Avenue, N.W. • Washington, D. C. 20418

NOTICE: The project that is the subject of this report was approved by the Governing Board of the National Research Council, whose members are drawn from the councils of the National Academy of Sciences, the National Academy of Engineering, and the Institute of Medicine. The members of the committee responsible for the report were chosen for their special competences and with regard for appropriate balance.

This report has been reviewed by a group other than the authors according to procedures approved by a Report Review Committee consisting of members of the National Academy of Sciences, the National Academy of Engineering, and the Institute of Medicine.

The National Academy of Sciences is a private, nonprofit, self-perpetuating society of distinguished scholars engaged in scientific and engineering research, dedicated to the furtherance of science and technology and to their use for the general welfare. Upon the authority of the charter granted to it by the Congress in 1863, the Academy has a mandate that requires it to advise the federal government on scientific and technical matters. Dr. Frank Press is president of the National Academy of Sciences.

The National Academy of Engineering was established in 1964, under the charter of the National Academy of Sciences, as a parallel organization of outstanding engineers. It is autonomous in its administration and in the selection of its members, sharing with the National Academy of Sciences the responsibility for advising the federal government. The National Academy of Engineering also sponsors engineering programs aimed at meeting national needs, encourages education and research, and recognizes the superior achievements of engineers. Dr. Robert M. White is president of the National Academy of Engineering.

The Institute of Medicine was established in 1970 by the National Academy of Sciences to secure the services of eminent members of appropriate professions in the examination of policy matters pertaining to the health of the public. The Institute acts under the responsibility given to the National Academy of Sciences by its congressional charter to be an adviser to the federal government and, upon its own initiative, to identify issues of medical care, research, and education. Dr. Samuel O. Thier is president of the Institute of Medicine.

The National Research Council was organized by the National Academy of Sciences in 1916 to associate the broad community of science and technology with the Academy's purposes of furthering knowledge and advising the federal government. Functioning in accordance with general policies determined by the Academy, the Council has become the principal operating agency of both the National Academy of Sciences and the National Academy of Engineering in providing services to the government, the public, and the scientific and engineering communities. The Council is administered jointly by both Academies and the Institute of Medicine. Dr. Frank Press and Dr. Robert M. White are chairman and vice chairman, respectively, of the National Research Council.

Support for this project was provided jointly by the National Science Foundation, the National Oceanic and Atmospheric Administration, the National Aeronautics and Space Administration, the Department of Agriculture, the Department of Defense, the Department of Energy, the Department of the Interior, the Department of Transportation, the Environmental Protection Agency, and the National Climate Program Office under Grant Number NA87-AA-D-CP041.

Readers are reminded that the opinions expressed in these proceedings are those of the individual participants and do not necessarily represent the consensus viewpoints of the Board on Atmospheric Sciences or the Committee on Global Change.

Library of Congress Cataloging-in-Publication Data

Ozone depletion, greenhouse gases, and climate change.
Proceedings of the Joint Symposium on Ozone Depletion, Greenhouse Gases, and Climate Change, held at the National Academy of Sciences, Mar. 23, 1988.
Includes bibliographies and index.

1. Stratospheric ozone—Reduction—Congresses. 2. Climatic changes—Congresses. 3. Greenhouse effect, Atmospheric—Congresses. I. National Research Council (U.S.). Board on Atmospheric Sciences and Climate. II. National Research Council (U.S.). Committee on Global Change. III. Joint Symposium on Ozone Depletion, Greenhouse Gases, and Climate Change (1988: National Academy of Sciences)
QC881.2.S8097 1988 551.6 88-31544
ISBN 0-309-03945-2

Printed in the United States of America

COMMITTEE ON GLOBAL CHANGE

BOARD ON ATMOSPHERIC SCIENCES AND CLIMATE

CLIMATE RESEARCH COMMITTEE

ROBERT E. DICKINSON, National Center for Atmospheric Research, *Chairman*
ERIC J. BARRON, Pennsylvania State University
INEZ Y. FUNG, Goddard Institute for Space Studies, National Aeronautics and Space Administration
ANTS LEETMAA, Climate Analysis Center, National Oceanic and Atmospheric Administration
JENNIFER A. LOGAN, Harvard University
JERRY D. MAHLMAN, Geophysical Fluid Dynamics Laboratory, National Oceanic and Atmospheric Administration
W. RICHARD PELTIER, University of Toronto
JORGE L. SARMIENTO, Princeton University
GRAEME L. STEPHENS, Colorado State University
KEVIN E. TRENBERTH, National Center for Atmospheric Research
JOHN E. WALSH, University of Illinois

JOHN S. PERRY, Staff Director
KENNETH H. BERGMAN, Staff Officer

COMMITTEE ON ATMOSPHERIC CHEMISTRY

ROBERT E. SIEVERS, University of Colorado, *Chairman*
WILLIAM L. CHAMEIDES, Georgia Institute of Technology
ROBERT A. DUCE, University of Rhode Island
DIETER H. EHHALT, Institut fur Chemie der Kernforschungsanlage Julich GmbH, Federal Republic of Germany
FRED D. FEHSENFELD, Aeronomy Laboratory, National Oceanic and Atmospheric Administration
ROBERT C. HARRISS, University of New Hampshire
CHARLES E. KOLB, Aerodyne Research, Inc.
RONALD G. PRINN, Massachusetts Institute of Technology
F. SHERWOOD ROWLAND, University of California, Irvine
STEPHEN E. SCHWARTZ, Brookhaven National Laboratory

JOHN S. PERRY, Staff Director
FRED D. WHITE, Staff Officer

COMMISSION ON PHYSICAL SCIENCES, MATHEMATICS, AND RESOURCES

Foreword

One motivation for organizing this symposium was the unusual convergence of a number of observations, both short term (less than 100 years) and long term (up to tens of thousand of years), that defy an integrated explanation. Of particular importance are surface temperature observations—especially the fact that 1987 was one of the warmest years on record globally—and observations of upper atmospheric temperatures, which have declined significantly in parts of the stratosphere. There has also been a dramatic decline in ozone concentration over Antarctica that was not predicted. Significant changes in precipitation that seem to be latitude dependent have occurred. There has been a threefold increase in methane in the last 100 years; this has been a problem in that a source does not appear to exist for methane of the right isotopic composition to explain the increase. The unexpectedly large methane increase provided the motivation for studying methane outgassing in the tundra permafrost regions. Thus, there are many climate and climate-related trends in progress that need study.

What is somewhat alarming is the relatively rapid change in climate, in some instances, that paleoclimatic data show. For example, a change well documented by Greenland ice core data is that, some 10,000 to 13,000 years ago, Greenland surface temperatures changed by about 5 or 6°C in a period of about 40 years. The explanation currently favored is that this change was due to a concurrent change in the mean position of the North Atlantic polar front. However, similar rapid changes are now being noted based on studies of ice core data from the Vostok, Antarctica, station. Hence, it seems that the comparatively rapid change in the North Atlantic area coincided with equally rapid change in the antarctic region.

Thus there are numerous exciting challenges in the area of climate change studies. One purpose of this symposium was to look at how all these indications fit together. Many of the problems we are seeing are associated with radiative gases being contributed to the atmosphere and their consequent effect on the climate, but there are complicated problems of atmospheric chemistry and climate interactions that need to be resolved and that served as the motivation for holding this symposium.

On behalf of the federal agencies, as represented by the National Climate Program Office, that are concerned with issues of climate and climatic change, I wish to thank the Board on Atmospheric Sciences and Climate and the Committee on Global Change of the National Research Council for organizing the symposium that resulted in these proceedings.

Alan D. Hecht, *Director*
National Climate Program Office

Preface

Global change is a topic that is of great importance to everyone. Nevertheless, it is often a nebulous concept in the minds of scientists from the individual disciplines involved. It is specific problems like the one addressed by this symposium, involving atmospheric dynamics, the coupled climate system, atmospheric chemistry, and the biota, that really indicate the need for a global change program. When specific problems of this nature arise, there seems to be little difficulty in getting scientists from different disciplinary backgrounds to work together with one another. The symposium reported on here can perhaps serve as a model for how such interdisciplinary, global change problems should be attacked.

In these Proceedings of the Joint Symposium on Ozone Depletion, Greenhouse Gases, and Climate Change, held at the National Academy of Sciences on March 23, 1988, a distinguished group of 10 experts address the important issues of stratospheric ozone depletion, possible linkages to increasing concentrations of greenhouse gases, and their combined effect in causing global climate change to occur. The suggestion for the symposium originated with the National Climate Program Office, National Oceanic and Atmospheric Administration, which asked the National Research Council's Board on Atmospheric Sciences and Climate to organize a meeting that would provide policymakers with sound scientific input on these issues. At an early stage, the planners realized that the symposium issues were of equally great concern to the planners of the International Geosphere-Biosphere Program (IGBP). Thus, the board joined with the Committee on Global Change (the National Research Council committee with responsibility for planning the U.S. component of IGBP) as well as with two of its own committees, the Committee on Atmospheric Chemistry and the Climate Research Committee, to plan and organize the symposium. The board and these committees hope that the result, as reported in these proceedings, is of value to specialists in the concerned disciplines and to the larger community of scientists, program managers, and decision makers as well.

On behalf of the board, I would like to thank the National Climate Program Office for its sponsorship of the symposium, the speakers whose contributions constitute these proceedings, and the several members of the National Research Council staff whose efforts made the symposium possible and resulted in these proceedings. Special thanks are due John Perry, staff director of the board, for his efforts in organizing the symposium and Kenneth Bergman, senior program officer of the board, for his work in preparing these proceedings.

Richard A. Anthes, *Chairman*
Board on Atmospheric Sciences and Climate

Contents

1
Introduction

Two important climatic issues—stratospheric ozone depletion and greenhouse gas increase—and the apparent connection between them led to the holding of this symposium. Theory predicts that ozone depletion should be occurring as a result of chlorofluorocarbons and halons in the stratosphere. Recent measurements confirm that such depletion is taking place on a global scale and is especially pronounced in the antarctic stratosphere. Global tropospheric warming due to increasing greenhouse gases has been an important climatic issue for many years, and several symposia and workshops have previously been held on this topic. However, recent data have made it increasingly apparent that the projected increase in greenhouse gases and the associated tropospheric warming and stratospheric cooling will indirectly affect ozone concentrations in the stratosphere, which in turn will affect tropospheric climatic conditions. Thus, the two issues are inextricably entwined and form part of the larger global change issue that recognizes that essentially all components of the earth-atmosphere-ocean-biosphere-cryosphere system interact with and affect one another, often in ways that are currently not well understood.

This symposium was primarily concerned with the linkages between ozone depletion and increasing greenhouse gases and with their combined effect in causing climate change to occur on a global scale. The presentations in these proceedings review the current state of knowledge about stratospheric ozone depletion, discuss the probable effect of predicted greenhouse gas increase on future ozone trends, summarize observational data on changing atmospheric chemistry and associated atmospheric temperatures, and describe the continuing effort to model and predict future scenarios of climatic change relative to ozone and greenhouse gases in both the stratosphere and the troposphere. Some of the questions and answers that followed the presentations have been included when they highlight noteworthy points that were not covered in the presentation itself. The request by the National Climate Program Office for a symposium on the above related issues is included as Appendix A, and the symposium agenda and participants are given in Appendix B. Appendix C is a glossary of special terms and abbreviations.

The first presentation, by William C. Clark, provides an overview of the global change issue and indicates the role of the symposium in furthering the goals of that larger effort. Daniel L. Albritton discusses the observational and analytical information on stratospheric ozone depletion that led to the signing of the September 1987 Montreal international agreement to restrict the input of ozone-destroying halocarbons into the atmosphere. Robert T. Watson's presentation describes the evidence for an antarctic stratospheric ozone "hole" and indicates that chlorine compounds are implicated as the primary cause of this phenomenon.

In his paper, F. Sherwood Rowland discusses the long-term outlook for stratospheric halocarbon concentrations and the associated outlook for ozone concentrations. He also presents evidence of recent stratospheric ozone depletion in the Northern Hemisphere, especially in winter. Mario J. Molina describes the specific heterogeneous chemical processes, believed to occur in the polar stratosphere, that result in the efficient destruction of ozone by free chlorine atoms and chlorine oxides. James G. Anderson discusses the less efficient homogeneous gas-phase catalytic process that destroys ozone elsewhere in the stratosphere. Anderson also presents additional evidence that shows why heterogeneous processes are especially efficient when ice clouds form in the polar stratospheric regions.

Jerry D. Mahlman reviews the combined effect of trace gases on changing stratospheric temperatures and circulation. Mahlman indicates that these stratospheric changes will feed back on the behavior of the trace gases and on the concentration of ozone, necessitating the use of dynamic modeling in order to fully understand and accurately predict the changes that will occur. Robert C. Harriss discusses the outlook for increased concentrations of one of the trace gases, methane, that affects stratospheric ozone chemistry and is also a tropospheric greenhouse gas. Harriss indicates that the extensive peat bogs and marshlands of the arctic slope are likely to be prime sources of increased atmospheric methane as the climate warms, providing a positive feedback for tropospheric warming and affecting stratospheric ozone distribution.

Kevin E. Trenberth reviews the evidence for global temperature trends, including both tropospheric warming and stratospheric cooling. Trenberth notes that the short duration of reliable stratospheric records along with other data problems makes it especially difficult to establish unambiguous temperature trends in the stratosphere. In the last paper, Robert E. Dickinson reviews the progress made with general circulation models in predicting the likely climatic changes engendered by increasing greenhouse gases and indicates those aspects of the climate system that are critical and in need of further model development.

In summary, the Joint Symposium on Ozone Depletion, Greenhouse Gases, and Climate Change reviewed the magnitude and causes of stratospheric ozone depletion and examined the connections that exist between this problem and the impending climate warming due to increasing greenhouse gases. The presentations of these proceedings indicate that the connections are real and important, and that the stratospheric ozone depletion and tropospheric greenhouse warming problems must be studied as parts of an interactive global system rather than as more or less unconnected events.

2
Global Change and the Changing Atmosphere

WILLIAM C. CLARK
Harvard University

A considerable set of long-term developments has put us in the somewhat awkward position today of having multiple programs and multiple problems that are partially overlapping but that lack well-understood linkages. One hundred and twenty years ago, the Italian geologist Antonio Stoppani wrote eloquently about the importance of understanding the connections, on a global scale, among human interventions, how the earth's surface was being transformed, and how that affected local climate and the atmosphere. Sixty years later, this theme was picked up by the "patron saint" of this gathering, Vladimir Vernadsky, in his lectures at the Sorbonne. He put forward—almost as a matter of religious faith, considering the lack of observations at the time—the notion that a fuller understanding of many of the atmospheric and related problems of the day would be served not by greater specialization in narrow subdisciplines but rather by trying to fashion a concept of the biosphere as a whole. Little research developed along these lines over the next 20 to 30 years because there were neither data nor instruments nor testable theories that would let anyone go beyond the assertion that the integrated, interdisciplinary perspective might be a useful approach.

Thirty years ago marked the occurrence of the International Geophysical Year (IGY), which for the first time gave us, on a systematic international basis, a worldwide perspective on environmental change in and of itself. It marked the start of the now infamous carbon dioxide monitoring series and of other such data records. It provided a window on the possible global-scale perturbations of many of the more important components of the geosphere-biosphere system by human activities.

In the years since the IGY, there have been broad advances in the basic sciences of the earth: biogeochemical processes and climate-biotic interactions, and how these both interact with human processes. The fundamental nature and importance of these advances are not diminished by the fact that many of them came about in response to specific problems of environmental change thought to offer threats to society. Examples of advances in atmospheric sciences include acid deposition programs, the stratospheric ozone programs to be discussed in this symposium, and climate change programs. Each of these programs, although building on and adding to a base of fundamental research, has pushed our understanding far deeper into areas of science linked to specific problems than into other areas that are every bit as interesting and every bit as fundamental to an integrated understanding of biosphere dynamics.

What is really new in the last 10 years is the pervasive conviction that the connections among the relatively well researched problem areas are not side effects but are central to our basic understanding of environmental change. In a sense, we are now saying that Vernadsky was right, that understanding particular changes in the geosphere-biosphere system, be they in climate, stratospheric chemistry, or atmosphere-soil interactions, requires a certain commitment to studying the interactions among those changes as well as the individual problems themselves.

Recognition of these interconnections is occurring in the scientific, administrative, and policy communities. Scientifically, we see it in the recognition that deforestation affects climate change, that climate change influences stratospheric chemistry involved in ozone depletion, and that the chlorofluorocarbons involved in ozone depletion feed back again on climate change. The neat separations between these areas that were almost complete 10 or even 5 years ago are simply not holding up today.

Administratively, we are seeing a scramble by the executive agencies to rearrange their budgeting criteria and evaluation mechanisms to support the kinds of research that deal with the connections. Sometimes it appears that administrative perspectives have forced scientists to look at the interactions among elements of global change.

More often, at least from the point of view of the science community, administrative structures have lagged behind scientific recognition of where the challenges for research lie. But nobody who is aware of what has been going on in the National Aeronautics and Space Administration (NASA), National Science Foundation (NSF), and the other executive agencies, even in the Office of Management and Budget, over the last several years, can help but realize that a determined, if imperfect, effort is under way to try to rearrange our funding priorities and our abilities to support long-term research efforts that address this notion of important connections among environmental changes.

Finally, a point that should not be overlooked is that this notion of connections is achieving more and more significance for management and policy. There is a growing recognition that, for example, we cannot set our policy toward fossil fuel use by looking at the greenhouse implications alone. One has to recognize that any policy change affecting fossil fuel use will also affect acid deposition, greenhouse warming, corrosivity of the atmosphere, and so on. This line of reasoning also applies to land-use management strategies, industrial policy, and similar policy issues. We have an opportunity to fashion arguments that people in the executive agencies and Congress can really use to try to advance policies in these very complex areas, by helping them to see these linkages and to explain them to their constituencies. The scientific community is beginning to recognize the opportunity but has done little so far to provide useful conceptual tools and means of communicating these linkages that can be used to build the social and political consensus necessary for action.

So, where is all this taking us? Many places, but most obviously to the International Geosphere-Biosphere Program (IGBP). IGBP has been brewing for many years out of a recognition of the existence of connections among disciplines. A couple of years ago in Berne, Switzerland, the International Council of Scientific Unions General Assembly anointed IGBP with the avowed goal of describing and understanding interactive physical and biological processes that regulate the total earth system and the unique environment that it provides for life, the changes that are occurring within this system, and the manner in which they are influenced by human activities. Now, that is a goal of basic research in the earth sciences that is difficult to disagree with. The problem that has preoccupied scientists and administrators over the last several years is how such a goal can be approached in practicable, doable steps that, at a minimum, do no harm to scientific research already under way. Steps that, in a more optimistic vein, promote some of the new long-term research, observations, and synthesis that are necessary to turn the notion of connections into a real revolution in our understanding of, and ability to cope with, global change in the geosphere-biosphere system.

The challenge of implementing the goals of IGBP hinges on two issues, which unfortunately have not always been distinguished. One is substantive, the other organizational. Substantively, the problem is to identify the few really "new start" programs of experimentation and observation that could make the most difference in our overall understanding of the interlinked earth science system. In doing that, the first requirement is one that is analogous to a principle accepted by the medical community: Do no harm to existing programs that are under way. We have to recognize that the opportunity for doing such harm is monumental if the exercise is not conducted with very close attention to what works already and therefore does not need fixing or extra coordinating.

Organizationally, there are equally strong imperatives and challenges. Again, a first requirement is to do no harm to organizational frameworks that, through years of evolution, are finally at the stage where they are supporting programs that are actually helping us to get on with the business of increasing understanding. Second, having ensured that we do as little harm as possible, we must make sure that the interdisciplinary linkages mentioned earlier do not fall between organizational stools. Third, we must take steps to ensure that the organizations we do have in place do not impede research that is crossing over their historical boundaries of self-definition. Finally, the ultimate challenge is to identify which, if any, new organizational frameworks would make a positive contribution to our ability to get on with the substantive work of understanding global change.

This brings us back to the purpose of this symposium. Obviously, a great many endeavors are under way to address both the substantive and organizational issues of global change. One of these is the recent report *Earth System Science, A Closer View* (National Aeronautics and Space Administration, Washington, D.C., January 1988) of the Earth Systems Science Committee. In addition to being the community's bid for the coffee table book of the year, this report represents an heroic effort to take an overview of the earth system, to identify some of the most important substantive problems, and to address the organizational difficulties of going after them effectively. It is a step in the right direction, but only if we build on it rather than simply setting it on the shelf. A second endeavor is the National Research Council's continuing review of geosphere-biosphere issues. Its most recent incarnation, the Committee on Global Change (CGC), is developing a preliminary plan for U.S. participation in the IGBP. The CGC is a noteworthy group in that it consists of not only atmospheric scientists and oceanographers but also biologists and geophysicists and even camouflaged sociologists. The challenge to the CGC is not to come out with simply another endorsement stating that linkages are important and that the geosphere-biosphere is out there and needs further study, but rather to really come to terms with the notion of significant, definable new problem areas for which solid research can be productive in the near term.

I remain puzzled as to just what role this symposium is going to play in moving on with the implementation of a research program on global change. But even that ambiguity is something that should be fostered. The symposium has clearly brought together a number of the best researchers in several closely related areas, in which just the type of connections that I referred to earlier are beginning to be unraveled and explored. Two National Research Council groups, the Board on Atmospheric Sciences and Climate and the Committee on Global Change, have joined forces to explore this issue. Finally, the symposium audience consists of most of the important Washington tribes: the Congress, the executive agencies, the nongovernmental organizations, and the press, all of whom are going to be necessary if a global change program is really to move forward and advance us some steps on the way to an understanding of the geosphere-biosphere system. I look forward to seeing what interactions—scientific, administrative, and political—this particular mix will provoke.

(In answer to a question): If the planned global change programs are as successful as they promise to be, they are going to create many more problems for the policy and management community than they solve, at least in the short run. They are going to turn up things that we did not know were going on and that we will be very uncertain and a little worried about. The global change program should not be viewed as a short-term response to existing, already known and understood problems. If a global change program is to have any long-term effect, the funding and support efforts also have to take a long-term perspective. We have to say over and over again to ourselves, to the agencies, to Congress, and to the public at large that the only way we will ever get out of playing "crisis response" to the degree that we have been doing of late (be it in ozone depletion, acid deposition, or some other "problem of the month") is to get the necessary broad-based basic research going. We need to take more initiative in showing how the halting progress we are making across the broad front of understanding really is improving our ability to deal with specific problems. The speed and efficiency with which efforts were mounted to take a look at the antarctic ozone phenomenon, once it was noticed, are an excellent illustration of how 10 to 15 years of basic preparatory research can prepare us to cope with surprises. We need to get that message across at least as much as we need to be concerned with getting the FY 1989 budget secured or getting a congressional hearing on immediate solutions to immediate problems.

(In answer to another question): Events will doubtless open up windows of opportunity for stepwise advances in support and understanding and will also close them on occasion. As a member of CGC, I feel that such opportunities will be wasted if we do not have on the table plans consistently and broadly supported by the entire community. These plans should be about very specific programs of measurement, experimentation, and documentation. They should not be broad statements saying where we hope to be, but instead plans specifying what we want to do next and exactly how we are going to do it. Then we can keep coming back and saying, "That is the sort of thing that should have been funded two years ago. That is the thing that should be supported now." We are not in that position now. All of the good works to date associated with global change—through NASA, NSF initiatives, and the National Research Council—have produced a good foundation but have not moved to a level allowing one to give congressional testimony and to talk to colleagues on the Hill and elsewhere and say, "Here is the research we are planning to do." Until we do that, the opportunities may come and go without our having a compelling rationale for pushing commitment and action.

3
Stratospheric Ozone Depletion: Global Processes

DANIEL L. ALBRITTON
Aeronomy Laboratory
National Oceanic and Atmospheric Administration

This talk summarizes the ozone science that led to the *Montreal Protocol on Substances that Deplete the Ozone Layer* (UNEP, 1987). It touches on three points: (1) what ozone theory said to those crafting the Montreal Protocol, (2) what ozone observations told that policy group, and (3) how policy responded to those science statements.

The Montreal Protocol represents a watershed in the way that science has interacted with policy and in the way that policy has responded to the science. The basic reason for my highlighting the science that led to the September 1987 Montreal meeting is that other speakers in this symposium will discuss what has been learned since that meeting was held. During the deliberations on the protocol structure, the antarctic ozone "hole" was discovered, and ozone scientists were wrestling with the question of its cause. Secondly, during the course of the Montreal debates, a scientific group of growing size and desperation were looking at the existing large data sets on ozone and sorting through what that mountain of conflicting data might mean. Because scientific consensus had not been reached, neither of these studies was put on the table as a rationale for the protocol. Now that the protocol has been concluded, one can ask the question: How well was that policy document crafted to incorporate the new science that has appeared in the last several months? This is an interesting question involving the interaction of science and policy.

The environmental issue associated with global ozone can be put in a nutshell: Man-made chlorine chemicals are depleting the stratospheric ozone layer. Atmospheric ozone is present in very small amounts in the lower atmosphere. However, it begins to increase in abundance at about 12-km elevation, marking the base of the stratosphere. Ozone reaches a maximum at around 25 km, which constitutes the center of the well-known ozone layer. This layer shields the earth's surface from biologically harmful solar ultraviolet radiation.

In 1974, two of today's speakers, Mario Molina and F. Sherwood Rowland, asked the question: What happens to the large volume of industrially produced chlorinated molecules that are released into the lower atmosphere, for which we know of no immediate atmospheric sinks? Their hypothesis as to the fate and consequences of these chemicals has five steps:

1. Man-made chlorinated compounds vastly exceed the natural ones.
2. The only loss of these compounds (mostly chlorofluorocarbons—CFCs) is through breakup by ultraviolet radiation in the stratosphere, where they are reduced to atomic components.
3. Chlorine and ozone can enter into a catalytic cycle whereby a chlorine fragment repeatedly destroys up to 10,000 ozone molecules before some other chemical process removes the fragment from the stratosphere.
4. The ozone layer is thinned by this ozone loss and hence passes more ultraviolet radiation to the surface.
5. Increased ultraviolet radiation at the surface is harmful to many of the surface biota, including humans.

Since 1974, this hypothesis has been improved and tested, and predictions have been made as to what the implications of the theory would be if we continue to release CFCs. Also, the ozone observational systems have been improved since 1974 by the use of both ground- and satellite-based instruments, so that the morphology of ozone is known in detail.

Based on the current understanding of theory, the science group at the Montreal meeting described the interactions of radiation, ozone, and chlorine to the policymakers in the following way:

1. The total amount of ozone overhead is a measure of the amount of ultraviolet radiation that is absorbed and hence does not

reach the surface. Surface ultraviolet radiation will increase if the overhead column of ozone is diminished.

2. Chlorine reactions deplete ozone in the higher part of the stratosphere, but feedback effects lead to a smaller ozone increase in the lower stratosphere, the net sum being a loss for the entire column.

3. Such a vertical redistribution of ozone would lead to local cooling and possible alteration of circulation patterns in the upper stratosphere. An increase at a lower altitude would lead to surface warming, because ozone acts as a greenhouse gas at lower altitudes.

4. The degree of predicted ozone loss varies with latitude. The greatest loss is predicted for higher latitudes.

The historical trend of the chlorine emissions predicted to cause the above effects is the following: There was a rapid increase from 1960 to 1974, a leveling-off and slight decrease to about 1983, and a renewed increase in the last few years. In 1974, the United States banned the use of CFCs in spray cans. As a result, there was a decline in subsequent CFC production because the use in other countries remained sufficiently low that its growth did not counterbalance the U.S. reduction. Thus, for a while, it appeared that the CFC-ozone problem might take care of itself. However, worldwide manufacture and use of these compounds have increased dramatically in recent years, leading to a renewed upswing in global CFC production at a rate of several percent a year. This renewed increase in CFC emissions was one of the main reasons that interest was rekindled in abatement regulations.

Three "what if?" emission scenarios were considered in describing future ozone responses:

1. The current increase of several percent per year continues.
2. The rate of increase is reduced by half.
3. CFC releases are frozen at the amount currently released annually.

What do scientists say about these scenarios in terms of the effect of both past and future CFC releases on stratospheric ozone? In terms of the global average ozone column, the advice to the policy group was that, if a freeze could be established in the near term, there would be a loss in global average column ozone of about 1 to 2 percent over the next 75 years. This scenario assumes that carbon dioxide and methane will continue to increase at their current rates. These two gases tend to offset the effects of atmospheric chlorine to some extent. Without their increasing presence, the ozone loss despite a CFC freeze policy would be considerably greater. In contrast to the effect achieved by a freeze, a continuing 3 percent annual growth rate would result in a loss of about 10 percent of the average column ozone in less than 75 years, assuming a continued increase in carbon dioxide and methane. This growth scenario and the projected loss of column ozone proved to be a very strong motivation for convening the Montreal Protocol.

Since the atmospheric retention time for most chlorine compounds is on the order of 100 years, stratospheric ozone levels would continue to drop in the near term even if all CFC releases were halted immediately. The long retention time also means that even a limited curtailment at the present time would be more effective in the long run than a more drastic curtailment later on. This knowledge also helped to bring about the protocol.

Another factor that the scientists described to the policymakers is the changes in the vertical profile of the ozone column in the event of a freeze. Atmospheric models predict that, even though the total column ozone would remain within a few percent of the present amount, the upper stratospheric ozone loss due to chlorine might be as large as 25 percent within the next 75 years. A 25 percent ozone loss at these altitudes implies a concomitant upper stratospheric cooling of about 5°C, which may alter stratospheric circulation patterns. (Natural variation of ozone at these levels is only about 3 percent.) The models also predict a 10 percent increase in ozone amount below 30 km. This would lead to a warming of the lower atmosphere and surface and would constitute a significant fraction of total surface and tropospheric warming that is predicted for all of the combined greenhouse gases.

Thus, even though a policy of a freeze in CFCs would minimize total column ozone loss, the predicted redistribution and consequent upper stratospheric cooling and tropospheric warming suggest that action to actually reduce the rates of CFC emissions would be more appropriate than a freeze.

Perhaps the most telling factor that scientists presented to the policymakers was the latitudinal dependence of column ozone depletion. With a freeze, the models predict that there would be less than a 1 percent loss of ozone at the equator, but they predict losses of 4 percent at 40°N and losses of as much as 7 percent at 60°N within the next 75 years. The higher-latitudinal values are well outside the range of natural variability of column ozone amounts. Therefore, a substantial reduction in chlorine emissions would be needed to reduce the predicted ozone loss at high latitudes to an amount similar to the natural variability.

In addition to the above theoretical considerations, observational data also influenced the policymakers at the Montreal meeting. Ozone is being monitored with three measurement systems:

1. Ground-based network of Dobson spectrophotometers. This network was set up in 1958 during the International Geophysical Year and consists of several dozen stations. The instruments are vertically oriented and measure total overhead ozone. The measurements show short-term fluctuations within plus or minus 2 percent. Regarding longer-term trends, the data show that ozone generally increased about 3 percent in the 1960s, remained roughly constant during the 1970s, and decreased about 4 percent in the 1980s.
2. "Umkehr" network of Dobson spectrophotometers. Instruments in this network obtain vertical profiles of ozone. These measurements show that, at the high altitudes above 30 km, ozone has declined irregularly by about 7.5 percent, with most of the decline occurring after 1980. This decline is on the order of what chlorine-ozone theory predicts for that period of time.
3. Solar backscatter ultraviolet (SBUV) satellite instrument. This instrument was launched in 1978. It obtains global coverage and also provides profile data that augment the limited measurements from the Umkehr ground measurements. These data show a large decrease in ozone of about 13 percent at the high latitudes above 30 km during the period of observation. A decline of this magnitude is *larger* than that predicted by the chlorine-ozone models for this time period.

Two somewhat opposed points of view about these observations emerged at the Montreal meeting. One group pointed out that the Umkehr and SBUV data showed depletions as a function of altitude and latitude that are in general agreement with the chlorine-ozone theory, but the magnitude of the depletion at the higher latitudes is even greater than predicted. The other group placed the greatest reliance on the Dobson instruments, noting that the Umkehr results are sensitive to the presence of volcanic dust, such as that from El Chichon, which erupted in 1982. They also pointed out that the SBUV sensors experience drift, for which it is hard to correct. Lastly, they suggested that the last 6 or 7 years is too short a period to establish a definite trend. Thus, there was no scientific consensus regarding the significance of the observational data.

As of September 1987, the understanding of the ozone issue could be summarized as follows:

1. The observations, although they suggested a decrease whose rough magnitude was similar to that predicted, were not considered entirely believable.

2. Theory, on the other hand, could justify some strong predictions:

a. If we do nothing about chlorine emissions, then it is likely that substantial ozone column losses will occur, particularly at high latitudes.

b. If we freeze emissions at 1985 rates, then global-average total-ozone column losses will likely be kept to an acceptable level, provided that carbon dioxide and methane increase as expected. However, there will be latitudinal and altitudinal variations that may prove unacceptable.

c. If we want to keep the latitudinal and altitudinal variations within acceptable limits in order to minimize high-latitude ultraviolet increase at the surface, surface temperature warming, and upper stratospheric cooling (with resultant circulation changes), then a substantial reduction in emissions will be necessary. (Here, "acceptable limits" means keeping the high-latitude column ozone loss and the high-altitude ozone loss to amounts no greater than those that result from natural variability, and the tropospheric-surface warming to less than one-fourth that expected from carbon dioxide.)

How did policymakers respond to this scientific input? Some highlights of the Montreal Protocol as it relates to this science follow. The scope of the protocol included all of the long-lived CFCs, as well as three commonly used halons, which are bromine compounds (these compounds cause ozone loss at a rate that is approximately ten times greater than that of the chlorine compounds). A timetable was established as follows:

- Entry in force—as early as 1989.
- 1990—Freeze CFCs at the 1986 levels.
- 1994—Cut emissions to 80 percent of 1986 levels.
- 1999—Cut emissions to 50 percent of 1986 levels.

The decreases to 80 percent in 1994 may constitute an approximate global freeze, in the sense that some countries will likely not participate in the protocol. However, the 1999 cut to 50 percent levels may be required. The answer lies in degree of participation and compliance, economics, new technologies, and demographics, all of which introduce considerable uncertainty in predicting the consequences of the protocol.

The protocol calls for automatic and periodic science reviews to allow for possible updating of its requirements. The signers will reconvene in 1990 to review the appropriateness of the protocol in the light of new observations and theory. A major international scientific review in 1989 will provide the science input for the 1990 meeting. By 1989, latitudinal effects should be better quantified by two-dimensional models. About 2.5 years of additional ozone data will help to resolve ozone trends. Also, more information on the mechanism of ozone depletion in the Antarctic, leading to the recent ozone hole phenomenon, will be available.

Most scientists involved with offering advice to the policymakers felt that a good match of policy decisions to the scientific information had been achieved by the protocol. The real foundation for the scientific issues presented at Montreal was the World Meteorological Organization's (WMO's) Global Ozone Research and Monitoring Project report (WMO-NASA, 1986). This report was clearly recognized by the policymakers as having three key attributes: (1) *authoritative*—the work of approximately 200 scientists is summarized therein, (2) *international*—the report represents the consensus evaluations of scientists from several countries, and (3) *comprehensive*—it covers not only ozone depletion but also its interaction with climate. The importance of this document in helping to establish the protocol indicates the crucial importance that the 1989 international scientific assessment will have. Plans for this effort are already under way by several groups, including the U.S. atmospheric agencies.

The watershed nature of the Montreal Protocol demands that we even improve on what scientists have been able to do in interacting with policymakers. While the stratospheric ozone and chlorine problem is extremely important in its own right, perhaps the most valuable lesson of the Montreal experience is an improved understanding of how the science and policy communities should interact in order to come up with a global action on a subject prior to the occurrence of unambiguously observed effects. With the greenhouse effect lying in wait for future scientists and policymakers, we need all of this kind of apprenticeship that we can get.

Question: Do the model predictions take account of atmospheric dynamics, or are they models of chemical reactions only?

Answer: The two-dimensional model predictions do take account of residual circulation effects. Admittedly, this is not a perfect representation, but it does explain the simultaneous behavior of some of the other trace gases.

Comment: Most people do not equate the Montreal Protocol with a true global freeze. Depending on the degree of participation or nonparticipation in the protocol, it may turn out that a true global freeze is not achieved until all the steps of the protocol timetable are completed by the participating countries.

Response: Certainly, this is a gray area. This brings up the need for studies by people who understand demographics, industrial responses to legislation, and compliance with past treaties of this sort, in order to come up with a data set with error bounds on the possible emission implications of the protocol as it stands. Such studies should be conducted by experts in these areas rather than by atmospheric scientists, who lack such expertise, and should be made available to scientists so that they can generate corresponding model predictions.

Question: Was there any consensus on the probable global impacts if nothing is done to control chlorine emissions for 75 years?

Answer: There were a number of algorithms developed by the Environmental Protection Agency that took a specified ozone decrease and translated that into certain effects. These were a part of the information provided to the policymakers in Montreal. These estimates of effects are hard to test, but they nevertheless provide some indication of likely impacts.

Question: In putting together the protocol, how much importance was given to the role that CFCs play in increasing the amount of climate warming induced by greenhouse gases?

Answer: The role of chlorine emissions in increasing the greenhouse gas effect was one of the motivating reasons for convening the Montreal meeting, although the primary motivation was the depletion of ozone at the higher latitudes. The greenhouse role of chlorine emissions is explicitly recognized by the protocol.

REFERENCES

United Nations Environment Program (UNEP). 1987. Montreal Protocol on Substances that Deplete the Ozone Layer. September 16, 1987. UNEP, Montreal.

World Meteorological Organization-National Aeronautics and Space Administration (WMO-NASA). 1986. Atmospheric Ozone, 1985: Assessment of Our Understanding of the Processes Controlling Its Present Distribution and Change. Global Ozone Research and Monitoring Project, Report No. 16, 3 vols., WMO, Geneva.

4
Stratospheric Ozone Depletion: Antarctic Processes

ROBERT T. WATSON
National Aeronautics and Space Administration

This paper describes the current understanding of the antarctic ozone hole phenomenon. A later speaker, James Anderson, will expand on the role of chlorine in causing the hole to occur. This talk addresses two questions: (1) What are the observations of ozone that define how large and "deep" the antarctic hole is? and (2) What is our present understanding of cause and effect? As Daniel Albritton noted, the Montreal Protocol did not explicitly take the appearance of the hole into account. However, the occurrence of the hole at about that time served as a major driving force to get the Europeans to view ozone as a serious issue and to get them to the table. The U.S. position at the Montreal meeting was based on the assumption that we did not know the cause of the antarctic ozone hole, although by then we recognized that an ozone hole existed. It is worth noting that the appearance of the ozone hole was an unexpected event in the sense that the models referred to by Albritton did *not* predict the hole. The ozone hole has made modelers realize that stratospheric modeling needs further work and has rekindled scientific interest in the problem of stratospheric ozone.

The ozone over Antarctica had, by October 1987, been reduced by more than 50 percent of its 1979 value (Watson et al., 1988; Figure 4-1). Locally, depletion was as great as 95 percent between 15 and 20 km altitude (Figure 4-2). Not only was the ozone level the lowest on record in 1987, but the seasonal period of depletion also lasted

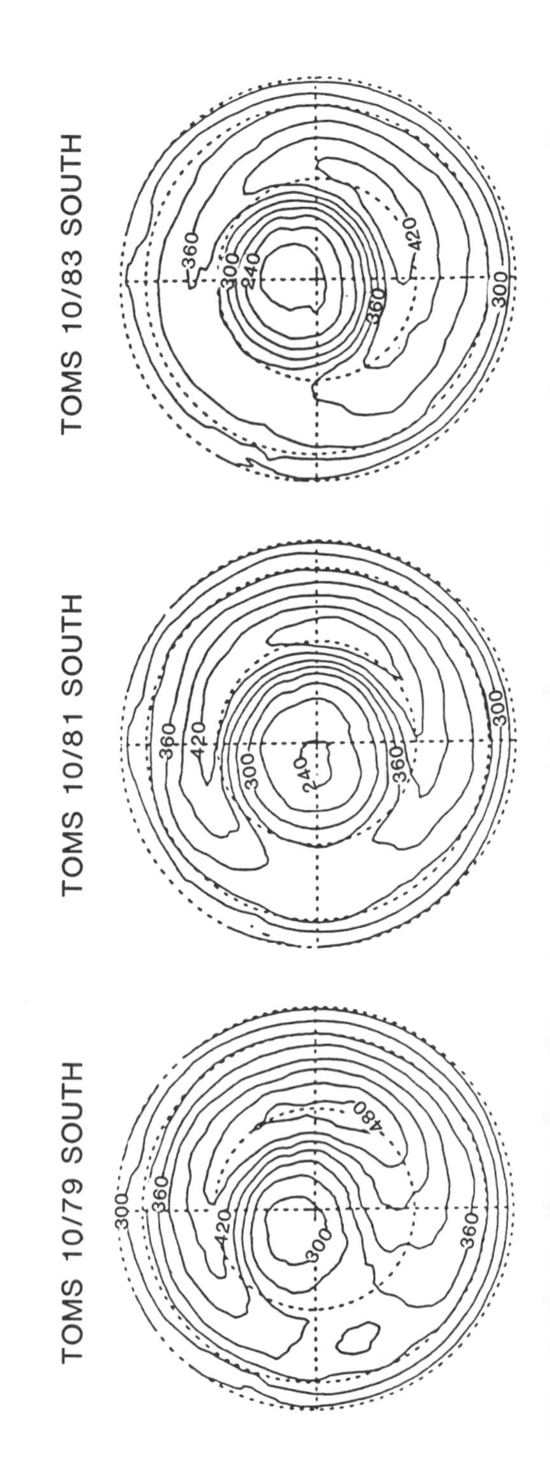
TOMS 10/79 SOUTH
300
360
420
300
480
360
TOMS 10/81 SOUTH
360
420
300
240
360
300
TOMS 10/83 SOUTH
360
300
240
360
420
300

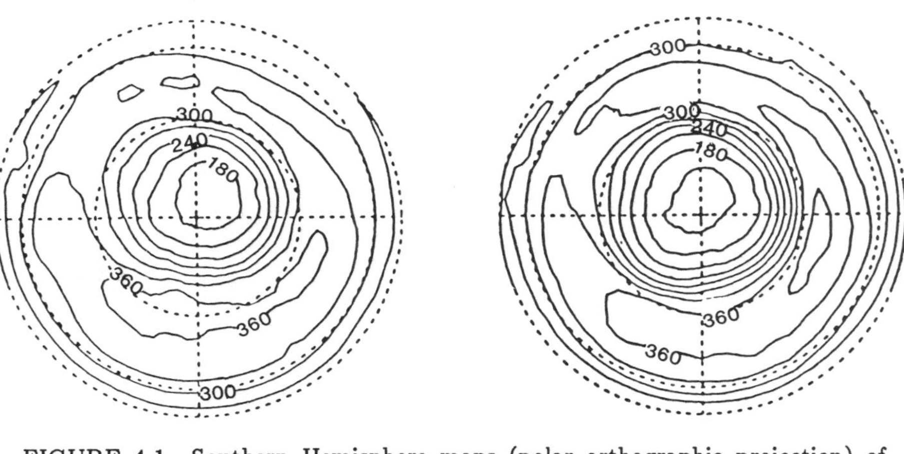

FIGURE 4-1 Southern Hemisphere maps (polar orthographic projection) of October monthly mean total ozone in odd years since 1979. Greenwich meridian is toward the top of each map. The equator is the outside edge.

the longest. The ozone hole itself did not fully disappear until late November or early December, 1987. This long-lasting minimum may have had significant consequences for the ecosystems in the antarctic region. The solar elevation angle is comparatively low by October, when the hole was at its deepest, but is much higher in November, when the ultraviolet (UV) effect might be stronger at the surface.

The ozone amount was also the lowest on record at all latitudes south of 60°S latitude in 1987. Furthermore, the occurrence of strong depletion was a year-long phenomenon south of 60°S and was not confined to the spring season as in preceding years, although the greatest depletion occurred during the Southern Hemisphere spring. Therefore it is no longer a relatively local, isolated effect (Figure 4-3).

For 1987, the measurements show little in the way of an ozone hole as late as August 17. After that, the ozone hole developed rapidly, especially after September 5, so that by October 5, the ozone over the middle of Antarctica had dropped from 320 Dobson units (DU) to 120 DU. The monthly average October mean of ozone decreased from about 300 DU in 1979 to 120 DU in 1987 over the middle of Antarctica. Additionally, the amount of ozone in the horseshoe-shaped maximum that extends out to at least 60°S decreased by around 100 DU in 1987, compared to 1979.

Two ozonesondes were obtained on October 6 and 9, 1987, at the U.S. Palmer Peninsula station (64°S latitude). This station is located near the edge of the region of low ozone. On October 6, when the edge of the strongly depleted region was poleward of the Palmer station, the ozone showed a fairly normal vertical profile. Three days later, the edge of the chemically disturbed and depleted region moved northward past the station, and the profile then showed a decrease of around 95 percent between 15 and 20 km. Other ozonesonde data from the South Pole, McMurdo, and Halley Bay stations, stations that were continually in the polar vortex region of depletion, show an almost complete disappearance of ozone after October 5. Hence, the ozone hole was a continent-wide phenomenon extending out to around the latitude of the Palmer station, where a steep horizontal gradient of ozone existed.

A series of Nimbus 7 Total Ozone Mapping Spectrometer (TOMS) satellite images shows some recovery of ozone in the antarctic hole by November 15, 1987, but the ozone amount remained below 175 DU over most of the continent (Figure 4-4a,b). By November 29, the minimum had moved from the polar region to over the Weddell Sea, surrounded by a large region of less than 200 DU. By about

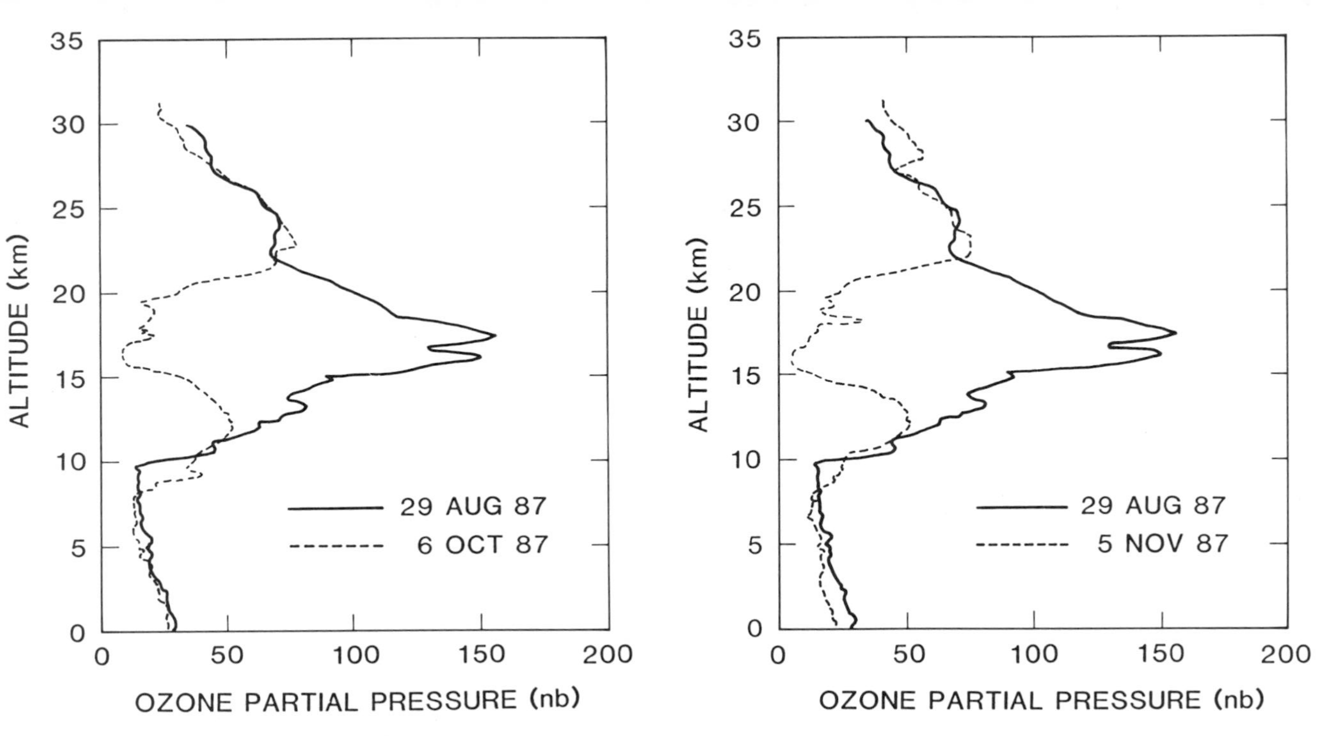

FIGURE 4-2 Vertical profiles of ozone using electrochemical ozonesondes from McMurdo station.

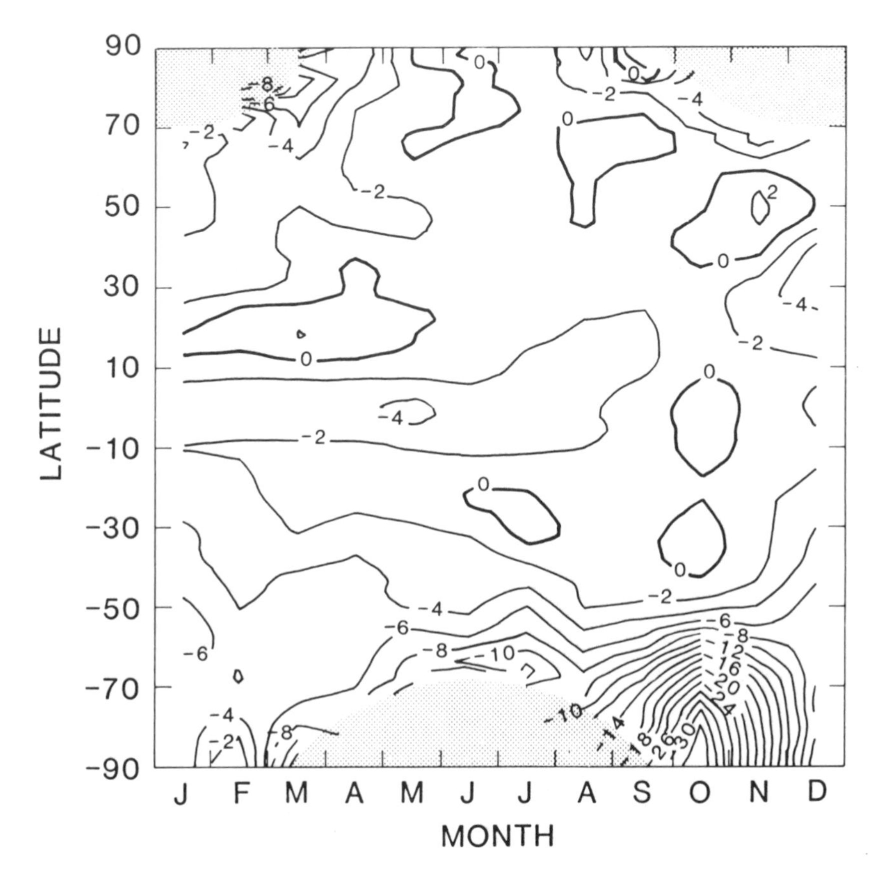

FIGURE 4-3 Percentage total column ozone changes, (1986-1987) minus (1979-1980), from the TOMS instrument as a function of latitude and month. Shaded areas are regions of no data during the polar night.

December 5, the hole had filled considerably, with a minimum of about 250 DU located well out over the Weddell Sea not far from the southern tip of South America. The relatively high sun angle of November and December coupled with this shift in the location of the ozone minimum likely resulted in a significant UV impact on the aquatic ecosystem of that region.

Further comparison of 1987 with earlier years indicates a progressively more rapid decrease of ozone during September in the later years. This fact is in agreement with a chemical hypothesis that, as more chlorine is added to the system, the rate of seasonal ozone decrease becomes greater.

A comparison of 1986 plus 1987 TOMS data minus 1979 plus 1980 data (2-year averages are used to remove possible effects of the quasi-biennial oscillation) shows that at least a 5 percent decrease in ozone occurred at all seasons south of about 50°S latitude. The satellite data have been corrected for drift relative to the ground-based stations and are believed to be correct to within 1 percent, but possibly 2 percent in the immediate South Polar region.

TOMS data from the Northern Hemisphere show a decrease in ozone from 1979 through 1985. This is consistent with an increase in trace gases, primarily chlorofluorocarbons (CFCs), and a decline in solar output (solar maximum in 1979-1980, minimum in 1985-1986). From 1985 to 1987, the ozone curve flattened or increased slightly, consistent with increased solar output after 1985 that countered the effects of increasing trace gases. In the Southern Hemisphere, however, the data show a continued decrease after the 1985 solar minimum (Figure 4-5). This is consistent with the concept that the antarctic ozone hole phenomenon causes a dilution effect throughout much of the Southern Hemisphere.

We now turn to the question of what is causing the antarctic ozone hole. The unique meteorology of the antarctic region in the winter and spring seasons results in the development of a strong polar vortex that consists of an air mass that is largely isolated from air farther north. Within the vortex, temperatures become cold enough to form stratospheric ice crystals. The ice crystals then allow unusual chemical interactions among nitrogen, hydrogen, and chlorine atoms. The weight of observational evidence indicates that the chlorine is primarily responsible for the ozone hole. Without chlorine in the antarctic stratosphere, there would be no ozone hole. (Here "hole" refers to a substantial reduction below the naturally occurring concentration of ozone over Antarctica.)

The relevant chemical reactions occur within the polar vortex. The vortex is not a uniform cylinder but has a shape that varies with altitude and is strongest and most isolated above the 400-K isentropic surface, around 15 km and above. Below 15 km, there is considerably more exchange of air between the midlatitudes and the polar regions. Hence, it is easier to explain behavior above 15 km in terms of chemistry than below 15 km, where atmospheric dynamics has an important role. Behavior below 15 km is still largely unexplained and is a matter of active research.

The National Aeronautics and Space Administration (NASA), the National Oceanic and Atmospheric Administration (NOAA),

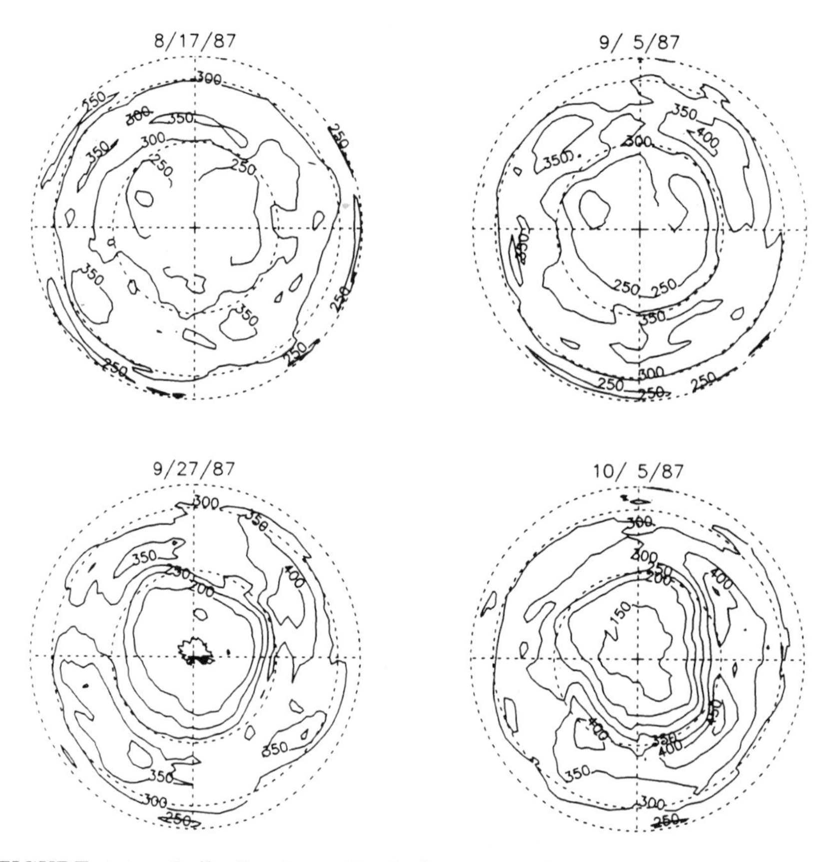

FIGURE 4-4a Daily Southern Hemisphere maps (polar orthographic projection) of total ozone measured by TOMS for the indicated dates. Period shown includes the formation of the ozone minimum in 1987.

the Chemical Manufacturers Association, and the National Science Foundation collaborated on ground-based expeditions to McMurdo in both 1986 and 1987 and provided extremely important results. These included observations of chlorine dioxide by Susan Solomon, which served as a very good indicator of perturbed chlorine chemistry, as well as measurements of very low nitrogen dioxide, showing that the atmosphere appears to be denitrified.

I participated in an aircraft expedition that was based in Puntas Arenas, Chile. We flew a DC-8 all over the antarctic continent during a 6-week period. We used remote sensors looking upward to obtain total column measurements of a wide variety of gases. We had a UV

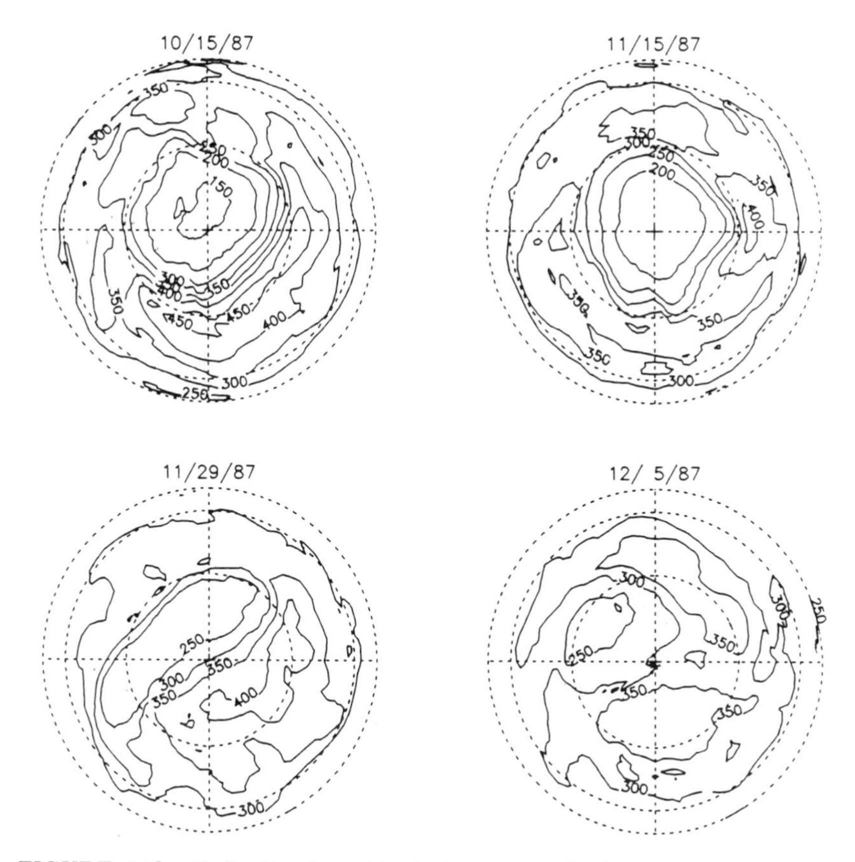

FIGURE 4-4b Daily Southern Hemisphere maps (polar orthographic projection) of total ozone measured by TOMS for the indicated dates. Period shown includes the breakup of the ozone minimum in 1987.

spectrometer similar to the one that Susan Solomon took to the ice to look at chlorine dioxide and bromine monoxide (BrO). We had a lidar instrument to look at both ozone and the aerosols. The ER-2 aircraft flew as high as 18.5 km in geometric altitude and measured several quantities: bromine and chlorine monoxide (James Anderson, Harvard University) to determine if the chlorine-bromine chemistry was perturbed, ozone (Walt Starr, NASA, and Mike Proffit, NOAA), total amount of nitrogen compounds (NO_y) in the atmosphere to determine whether they were enhanced or depleted (David Fahey, NOAA), whole air samples for CFCs, methane (CH_4), and nitrous oxide (N_2O) (Leroy Heidt, National Center for Atmospheric Research

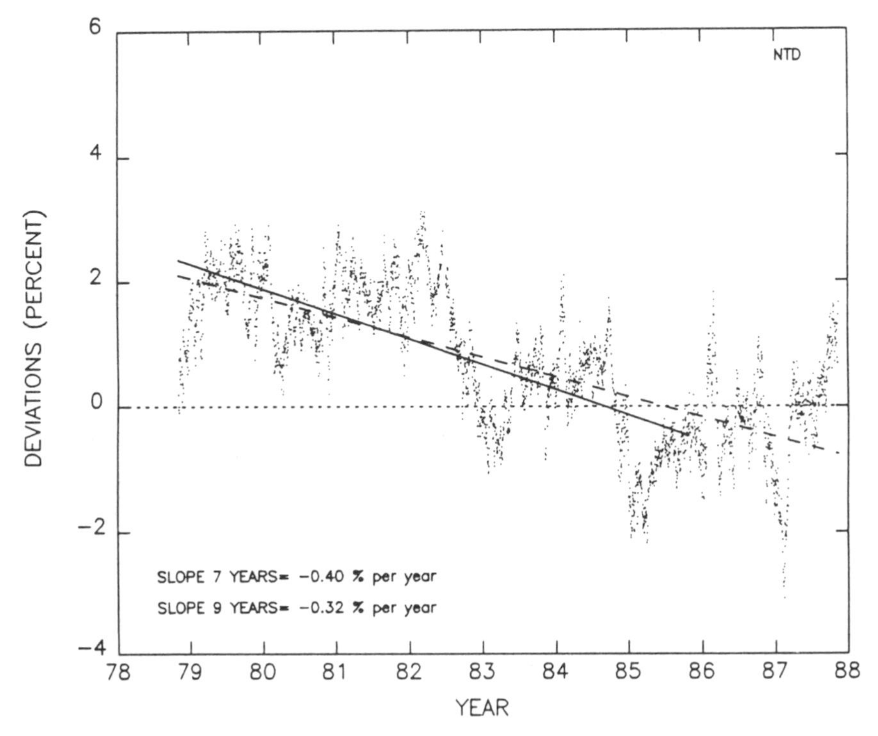

FIGURE 4-5 Total column ozone integrated from 53°S to 53°N latitude, from 1979 to 1987, as determined from TOMS data, shown as percent deviation from an arbitrary reference level.

(NCAR)) to determine rising or sinking motions in the air mass, and a large number of aerosol measurements, as well as a system to measure the temperature lapse rate (Bruce Gary, Jet Propulsion Laboratory (JPL)).

The data from this ensemble of instruments are being used to test the various hypotheses that have been proposed. These theories include the solar theory, whereby periodically the amount of nitrogen compounds is enhanced. These enhanced levels can catalytically destroy ozone in the lower stratosphere. This theory, if correct, implies that levels of oxides of nitrogen should be elevated; besides, the ozone hole should occur cyclically. The data showed that this theory is completely and utterly wrong. The oxides of nitrogen were measured as being unusually low. Some other theories that require an increase in nitrogen compounds are likewise incorrect.

The fluorocarbon-halon theory suggests that there should be a change in the partitioning of chlorine from the inactive forms of chlorine, namely hydrochloric acid and chlorine nitrate, into the active forms of chlorine, namely chlorine atoms and chlorine oxide radicals. Therefore, Anderson's tests were critical for determining whether the chlorine oxide (and bromine oxide) abundances were enhanced.

Another theory, advanced by K.-K. Tung, requires a change from downward to upward motions over Antarctica in association with other circulation changes. If this is correct, one should see enhanced levels of tropospheric trace gases such as nitrous oxide and methane in the lower and middle stratosphere. Therefore, the Heidt measurements were critical for this purpose.

The ER-2 aircraft could not climb higher than 18.5 km because of the very cold, dense atmosphere and the need to carry a lot of fuel for safety reasons. Also, it did not range farther south than 72°S latitude, about midway down the Palmer Peninsula. Therefore, many of the measurements were made close to the inside edge of the polar vortex. It would have been scientifically desirable to have flown higher and farther southward in the vortex. In any event, our flights from Puntas Arenas to 72°S and back were useful for comparing conditions just inside the vortex to those outside.

One of our first flights was made on August 23, 1987. Water vapor dropped from about 3 ppm outside the vortex to about half this value inside, indicating that the atmosphere inside the vortex was dehydrated. Ozone changes were only slight across the vortex boundary. However, the abundance of the chlorine monoxide radical (ClO) increased from about 10 parts per trillion by volume to about 500 parts per trillion. The nitrogen compounds (except for nitrous oxide) dropped from 8 to 10 parts per billion by volume (ppbv) to only 1.5 to 2 ppbv. Thus, the vortex atmosphere on this date was dehydrated, denitrified, and highly enriched in chlorine oxides, but with little effect on ozone levels.

By the end of our mission on September 22, the polar vortex atmosphere was still dehydrated and denitrified, the chlorine oxides had increased to about 1 ppbv, and the ozone concentration had dropped to less than half of its value outside the vortex. Therefore, it appears that a significant amount of time, approximately 1 month, is required for the chlorine oxides to destroy ozone. The concentrations of bromine oxides within the vortex were in the range of 3 to 5 parts per trillion on all flights. These small concentrations, compared to the chlorine oxide concentrations, imply that whereas chlorine oxides play a major role in destroying stratospheric ozone by the ClO-ClO mechanism, the proposed ClO-BrO mechanism for destroying ozone is of relatively minor importance, accounting for less than 10 percent of the total ozone depletion.

The DC-8 flew nearly to the South Pole and obtained many bulk column measurements. These show that, in crossing inside the polar vortex, nitrogen dioxide drops off significantly, nitric acid peaks around 70°S and then drops, chlorine nitrate reaches a maximum at the edge of the vortex, and hydrochloric acid falls off significantly within the vortex. The JPL and NCAR measurements are in excellent agreement with each other. The chlorine nitrate results from the presence of both ClO and nitrogen dioxide; hence it maximizes near the vortex boundary, where ClO is increasing rapidly but nitrogen dioxide is decreasing rapidly.

Calculations by Anderson show that ozone depletion at the 410- and 420-K isentropic surfaces between August 23 and September 22 can be almost entirely explained by the amount of ClO present if one assumes that the ClO-ClO mechanism is effective. At the 360-K surface, the calculated ozone loss is somewhat less than the observed loss. At least we can say that above about the 400-K level, there does seem to be enough ClO to explain the observed ozone loss.

A number of measurements were obtained of particles, ranging from sulfuric acid particles (less than 0.1 micron) through nitric acid trihydrate (0.5 micron) to relatively large ice crystals (2 to 3 microns). These measurements tend to support the current hypothesis of how chlorine oxide concentrations become enhanced in the polar stratosphere.

Measurements of nitrous oxide and methane obtained at an altitude of about 18 km and near the inner edge of the vortex did not give any evidence of upward vertical motions. Since the ER-2 flights did not penetrate poleward of 72°S, one cannot make a blanket statement that upward motions did not occur anywhere within the polar vortex region. However, these data, in conjunction with the other data, suggest that upward vertical motions do not play an important role in the ozone depletion process.

In summary, chlorine is intimately involved in the depletion of ozone, and most of the ozone loss can be explained quantitatively on a chemical basis. All theories, especially the solar theory, that require elevated concentrations of oxides of nitrogen are incorrect, and the apparent absence of large-scale upward motions suggests that the K.-K. Tung type of theory is wrong as well.

(In answer to a question about ice crystals and temperature): Since 1984, there has been an increase in the persistence of polar stratospheric clouds (PSCs) at 16, 18, and 20 km, with the PSCs persisting through October. This is possibly consistent with the temperature getting colder through October. We have looked at the temperature record, and there is no evidence for a change in stratospheric temperature from 1979 to 1987 in August or September, when the hole forms, but the stratosphere appears to be about 8°C colder in October and November at the 100-mb level over Antarctica. This implies that the temperature change is a result of ozone depletion rather than a cause of it. Since it is colder in October than formerly, PSCs seem to be persisting longer now.

Question: What would be the first signs of damage to the biota in Antarctica from increased UV radiation?

Answer: Presumably, the first sign would be a die-off of the phytoplankton and then the krill in the surrounding waters. Unfortunately, there are no long-term records of the phytoplankton population over the last 20 to 40 years, so a good comparison cannot be made. Laboratory studies suggest that enhanced levels of UV would be quite catastrophic to the phytoplankton and krill life in the region, but such measurements may not properly represent how the natural system works.

Question: How effective will the Montreal Protocol be in reducing the severity of the antarctic ozone hole?

Answer: There was no antarctic ozone hole from 1965 to 1970 with chlorine at a concentration of 2 ppbv. There is a huge antarctic ozone hole today with chlorine at 3 ppbv, and there is evidence that the ozone hole is enlarging and spreading. Under the Montreal Protocol, the concentration of chlorine will certainly rise to at least 5 ppbv and possibly to as high as 8 or 9 ppbv. Therefore, I believe that the protocol will do absolutely nothing to protect the antarctic region. The ozone hole may get worse, and there will be more hemispheric, and possibly global, ramifications. If policymakers believe that we should protect ozone over Antarctica, then it is quite clear that the Montreal Protocol will have to be revised and the measures made much more stringent.

Question: How is the edge of the southern polar vortex defined?

Answer: I have heard two different definitions of the vortex. They are based on the location of the steepest potential vorticity gradient and the location of the jet of maximum winds, which is about 5 to 8° of latitude wide. Some define the inner edge of the wind jet as the limit of the vortex air. This appeared to correspond closely to the boundary of the chemically perturbed region, and we usually encountered it around 68 to 70°S latitude. The alternate definition is the equatorward side of the belt of maximum winds, which extends to about 60°S. This definition is not consistent with our measurements. A related question is: How fast does air move across the polar vortex boundary? Several scientists are looking at this question.

Question: At lower latitudes where there are no stratospheric ice crystals, is the role of ice mimicked by other aerosols such as volcanic dust?

Answer: The next two papers address that question. The evidence that I have seen from laboratory studies indicates that liquid sulfuric acid particles will not provide such an efficient surface for heterogeneous chemistry, partly because the rate of reaction proceeds more slowly compared to that with ice crystals, and partly because the typical density of the sulfuric acid aerosols is less than that for ice crystals over Antarctica. Therefore, the likelihood of significant heterogeneous chemistry appears to be less at lower latitudes.

REFERENCE

Watson, R.T., M.J. Prather, and M.J. Kurylo. 1988. Present State of Knowledge of the Upper Atmosphere 1988: An Assessment Report. NASA Reference Publication No. 1208. National Aeronautics and Space Administration, Washington, D.C.

5
The Role of Halocarbons in Stratospheric Ozone Depletion

F. SHERWOOD ROWLAND
University of California, Irvine

This presentation will cover two topics: (1) halocarbons in the atmosphere and (2) the measurement of ozone.

Starting in 1978, my research group made gas chromatographic measurements of trichlorofluoromethane (CCl_3F, known as CFC-11) with air samples from many locations in both hemispheres that were judged to be sufficiently remote from local emission sources. Other research groups had collected similar data beginning as early as 1970. A set of our measurements of CCl_3F from 1979 shows only a small hemispheric difference in the lower atmosphere (Figure 5-1). Although about 95 percent of the chlorofluorocarbons (CFCs) are released in the Northern Hemisphere, the redistribution between the hemispheres is rapid enough that the Southern Hemisphere lags behind the Northern by only about 10 percent. Measurements from later summers show an increase at all latitudes. As part of the Global Atmospheric Gas Experiment (GAGE), intensive measurements have been made with automatic gas chromatographs operating at five stations most of the time since July 1978. These GAGE measurements through 1983 for dichlorodifluoromethane (CCl_2F_2, known as CFC-12) in Ireland and Tasmania (Cunnold et al., 1986) are shown in Figure 5-2, together with flask data, from the Oregon Graduate Center, for January measurements in Oregon and at the South Pole (Rasmussen and Khalil, 1986). All of the data show the level of

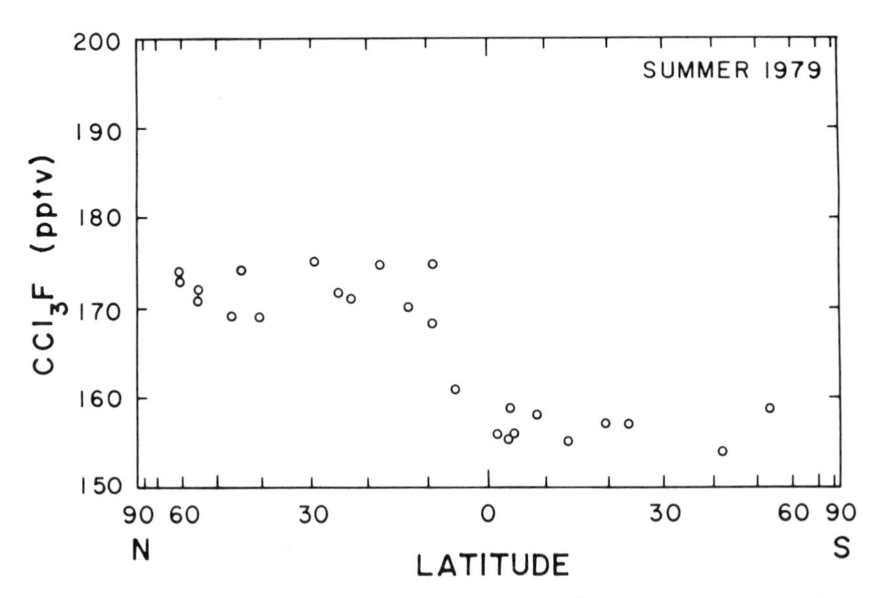

FIGURE 5-1 Tropospheric concentration (in pptv) of trichlorofluoromethane (CCl_3F) as a function of latitude for summer 1979. (Note nonzero origin of ordinate.)

CFCs rising rapidly during the 1980s, with the Southern Hemisphere lagging behind the Northern Hemisphere by about 10 percent.

In order to estimate the average lifetime of these molecules in the atmosphere, one needs to know the amount being put into the atmosphere for comparison with the amount that is still there. For CCl_2F_2, the estimate of its mean life is more than 100 years, and this is believed to be a fairly typical value for CFCs. Since the average CFC molecule has been in the atmosphere only about 10 to 12 years, 90 percent of them are still there. The estimated mean life for CCl_3F is given as 75 years.

We ran a numerical calculation in which CCl_3F release was assumed to grow exponentially until 1976, remain constant for 15 years until 1991, and then drop to zero after that. At 75 years after 1991, the amount of chlorine compound remaining in the stratosphere had dropped to 37 percent and at 150 years (i.e., 2160 A.D.) had declined to 13 percent of the 1991 amount. Most of the other CFCs have somewhat longer lifetimes, up to 140 years, so that an appreciable fraction may still remain in the atmosphere even after 200 years, given the above scenario. All indications are that the chlorine atoms

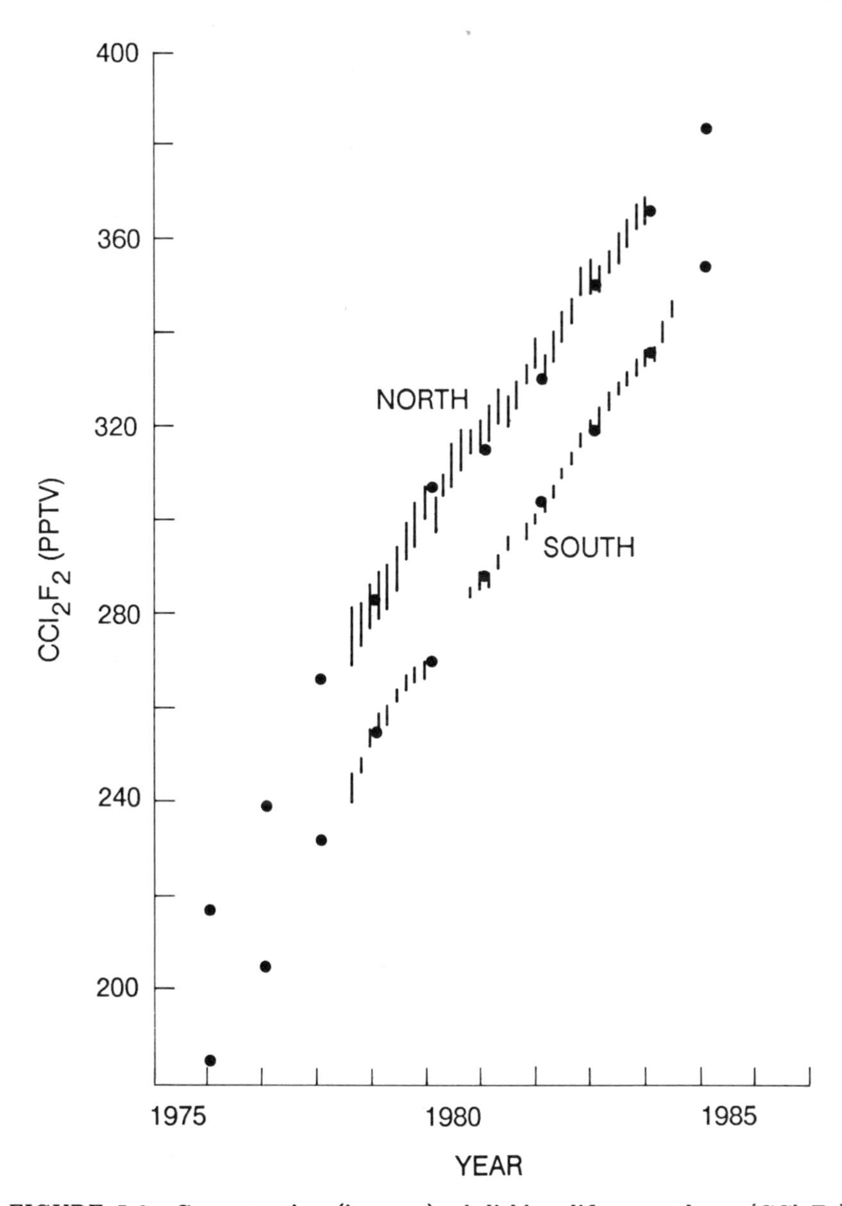

FIGURE 5-2 Concentration (in pptv) of dichlorodifluoromethane (CCl_2F_2) from 1978 to 1985 for GAGE measurements in Ireland (vertical lines labeled "North") and Tasmania (vertical lines labeled "South") (Cunnold et al., 1986), together with January flask data obtained in Oregon (upper dots) and at the South Pole (lower dots) (Rasmussen and Khalil, 1986). (Note nonzero origin of ordinate.)

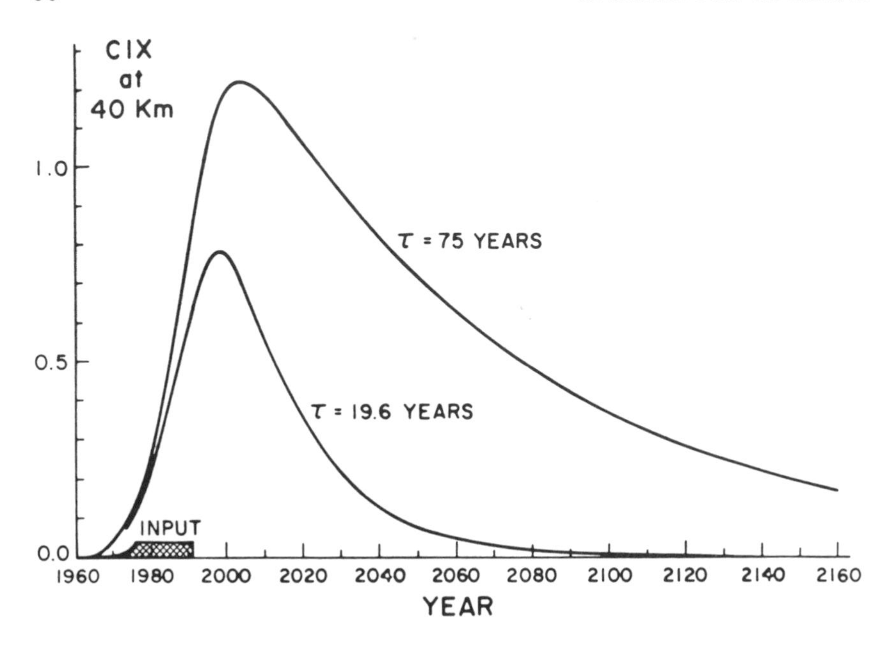

FIGURE 5-3 Concentrations (in ppbv) as a function of year of two typical stratospheric chlorinated molecules, one with a mean life of 75 years and the other with a mean life of 19.6 years. These concentrations are based on an assumed input of CFCs to the atmosphere (increasing to 1976, then constant to 1991, then no further emissions) as shown in the lower left of the figure. (Adapted from Rowland and Molina, 1976.)

released from CFCs are going to be with us for a long time, even if release of CFCs is discontinued tomorrow.

A related point concerns what the Montreal Protocol calls the "ozone depletion potential." Consider a chlorinated molecule with a mean life of 75 years and another with a mean life of 20 years (Figure 5-3). If their respective effects on ozone depletion are compared as a function of time, the difference between them does not become large until more than 50 years have passed (Rowland and Molina, 1976), by which time the 20-year compound will be largely gone.

If CFC emissions are assumed to continue into the future at a constant rate, stratospheric chlorine compounds will continue to increase (as shown in Figure 5-3). In 1974, the Northern Hemisphere contained chlorine compounds at about 1.8 ppbv; this has now increased to about 3.5 ppbv. Continued release at 1986 rates will result in increases to over 5.0 ppbv by the year 2000 (Figure 5-4). If, instead, we assume a 20 percent reduction of CFC emissions in 1994

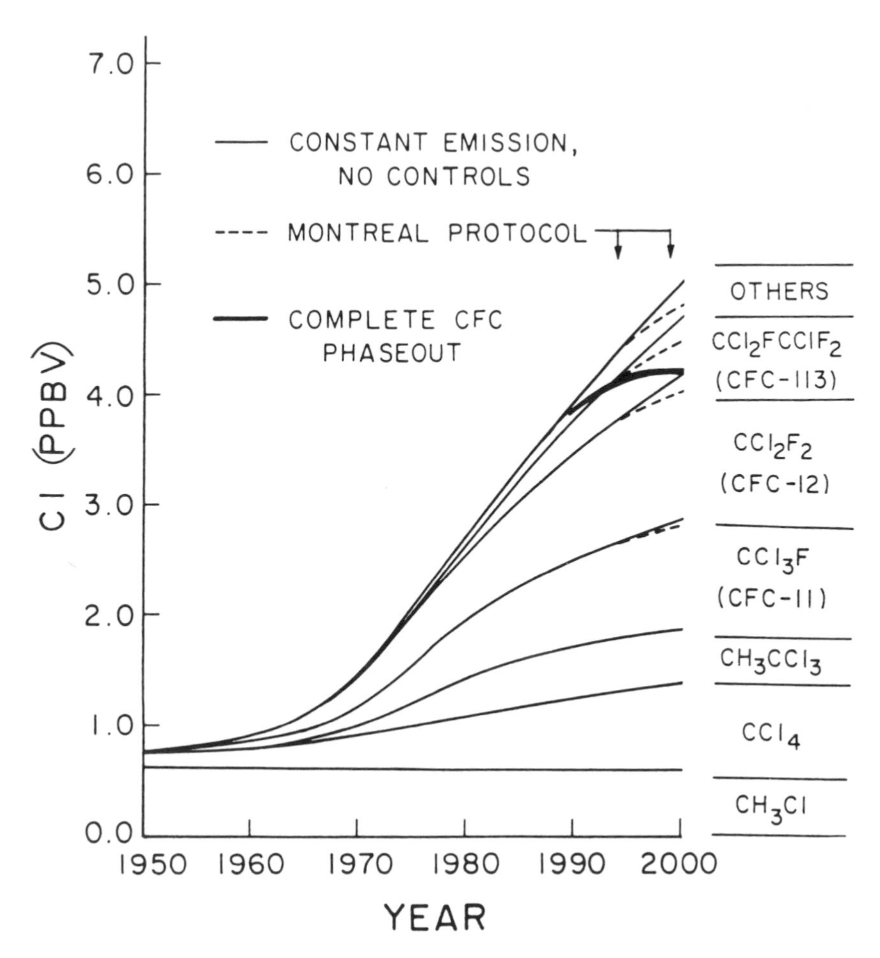

FIGURE 5-4 Increases in concentrations (in ppbv) of stratospheric chlorinated molecules assuming (1) continued release of CFCs at 1986 rates (solid curves), (2) a 20 percent reduction in release rates in 1994 and an additional reduction of 30 percent in 1999 (dashed curves), and (3) a complete phaseout of CFC emissions over a 10-year period beginning in 1989 (heavy solid curve).

and a further 30 percent reduction in 1999, as the Montreal Protocol requires, the predicted change in time will be different, assuming that all countries obey the protocol. Finally, if we assume a complete phaseout of CFC emissions over a 10-year period beginning in 1989,

TABLE 5-1 Trends in Total Ozone

	Percent Change	Reference
1970-1978	+0.28 ± 0.67	Reinsel et al., 1981
1970-1979	+1.5 ± 0.5	St. John et al., 1982
1970-1979	+0.1 ± 0.55	Bloomfield et al., 1983
1970-1983	−0.003 ± 1.12 per decade	Reinsel et al., 1984
	(−0.14 ± 1.08) per decade with sunspot series in model	

Source: World Meteorological Organization-National Aeronautics and Space Administration, Washington, D.C. (1986).

the amount of stratospheric chlorine compounds will still continue to increase for some decades.

The second topic addressed in this presentation is ozone measurements. The WMO-NASA report *Atmospheric Ozone 1985* (WMO-NASA, 1986) summarized recent calculations of trends in total global ozone. The trends (through 1983) are only slightly, if at all, downward (Table 5-1). However, the trend estimates given in the WMO-NASA report are misleading, as explained below.

The Arosa, Switzerland, station has the longest record of ground-based measurements of total ozone, made since 1931 with a Dobson spectrometer. Arosa is in the Swiss Alps at 47°N latitude. Neil Harris, University of California at Irvine, has examined the monthly averages of the data taken daily at Arosa. The amount of stratospheric ozone at Arosa, and generally in the north temperate zone, varies seasonally, with a peak in March or April and a minimum in October or November. The standard deviation of the measurements is very large during the winter season and comparatively small in the summer (Figure 5-5). If one compares the data for each month for the periods 1931 to 1969 and 1970 to 1986, one sees that there is less ozone, on the average, during the winter months in the later period (Figure 5-6). The greatest difference is observed for the month of December, with a loss that substantially exceeds the standard deviation for the data.

Measurements at Bismarck, North Dakota, which is also at 47°N latitude, began in 1963. We formed two sets of data, each 11 years long, one for 1965 through 1975 and the other for 1976 through 1986. A length of 11 years was chosen to permit comparison over two successive solar cycles. There is a wintertime loss of ozone during the

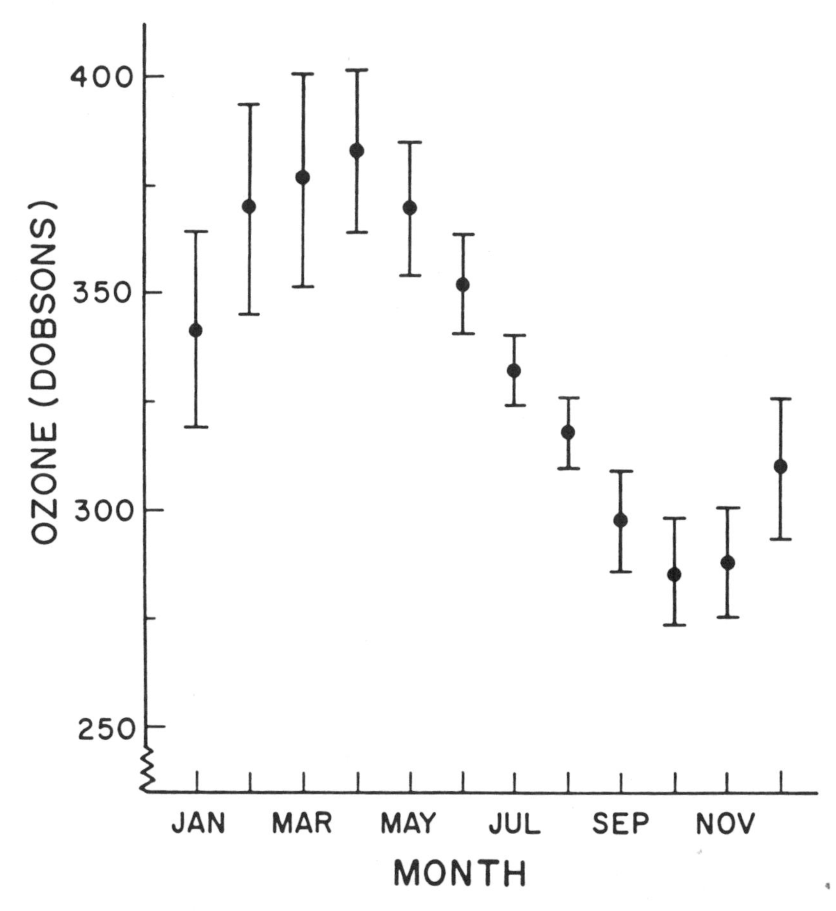

FIGURE 5-5 Mean monthly total column ozone measurements (Dobson units; includes data from Dütsch, 1984) at Arosa, Switzerland, for the period August 1931 through July 1986. Standard deviations from the mean of individual monthly data are also shown.

second period at Bismarck, similar to the loss reflected in the Arosa results. The wintertime loss also shows up in data from Caribou, Maine, which is also at 47°N. Hence, a wintertime depletion of ozone in the last decade or so has been observed for numerous northern stations.

Numerous factors are believed to affect the concentration of stratospheric ozone. The solar sunspot cycle affects the ozone concentration because there is increased UV radiation at around 200 nm during the sunspot maximum. Radiation of this wavelength can split

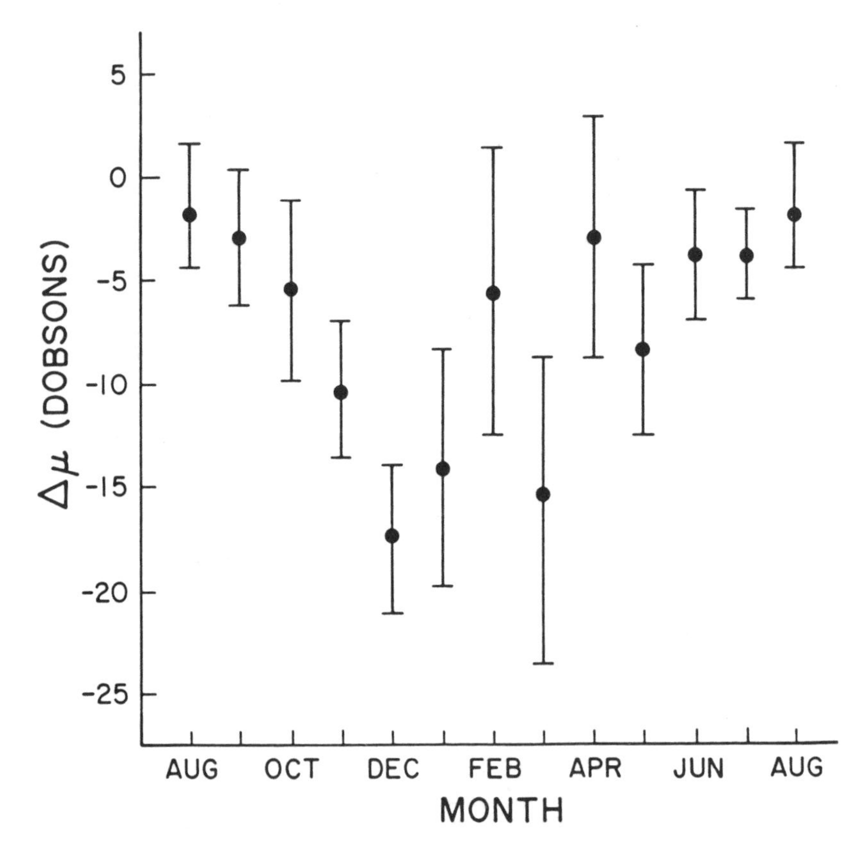

FIGURE 5-6 Mean and standard deviation of the *difference* in total column ozone measurements (Dobson units) between the periods August 1931 to December 1969 and January 1970 to July 1986. Negative values indicate lower ozone concentrations during the later period.

an oxygen molecule into two oxygen atoms, each of which can attach to another oxygen molecule to make a molecule of ozone (O_3). Hence one expects a cycle of increasing and decreasing ozone that closely parallels the sunspot cycle itself.

Another factor that affects ozone is the "quasi-biennial oscillation" (QBO), with a period of 26 or 27 months. The QBO is a cycle of varying wind directions in the equatorial stratosphere that can influence the flow of ozone to the north or the south in different phases of the cycle. The QBO shows up clearly in data from Varanasi and New Delhi, India, as a regular variation with the QBO frequency of about 26 months. Unfortunately, the records of these two Indian stations, which correlate so well after 1977, cross one another earlier in a manner suggesting a problem with the continuing absolute calibrations at one or both stations. Long-term calibration problems such as these make the data at some stations not very useful for the determination of trends in ozone concentrations over a 20- or 30-year period.

In order to determine long-term ozone trends, Rumen Bojkov from the Atmospheric Environment Service in Canada, Peter Bloomfield, a statistician from North Carolina State University, Neil Harris, and I have compiled data from all the Dobson stations and the Soviet stations by latitude bands. (The Soviet stations use a different instrument, the M-83, with somewhat different characteristics but qualitatively similar measurement techniques.) The data span the period 1965 to 1986 and are reported in terms of a "ramp" fit to the 22 years of data. The statistical analysis includes variations from the solar cycle and the QBO, plus an assumed linear change after 1969 from an otherwise constant value from 1965 to 1969. The data have been recorded as percentage changes over the 17-year period 1969 to 1986, from the linear ramp coefficients.

The results, reported by the Ozone Trends Panel (Watson et al., 1988), were as follows, on a monthly basis:

1. Between 53°N and 64°N (Figure 5-7a): not much change in July to September but very substantial decreases in December to March, similar to the Arosa data results. The QBO had a 2 percent variation in the statistical analysis, and the solar cycle showed 1.8 percent more ozone at the solar maximum than at the solar minimum. (These variations were removed from the data in order to study the long-term trends.)
2. Between 40°N and 52°N (Figure 5-7b): again, a marked difference between summer and winter trends. The QBO and solar cycle are again apparent.
3. Between 30°N and 39°N (Figure 5-7c): less difference between summer and winter trends, but all months show decreases in ozone, some of them large enough to be statistically significant, including decreases observed for July.

Regression coefficients were also calculated including successive years of data from 1965 through 1980, 1981, 1982, and so on, up through 1986. The coefficients show some variation with the additional years of data for the 53°N to 64°N zone and for the 30°N to 39°N zone, but none of the changes appears to be statistically significant. A negative

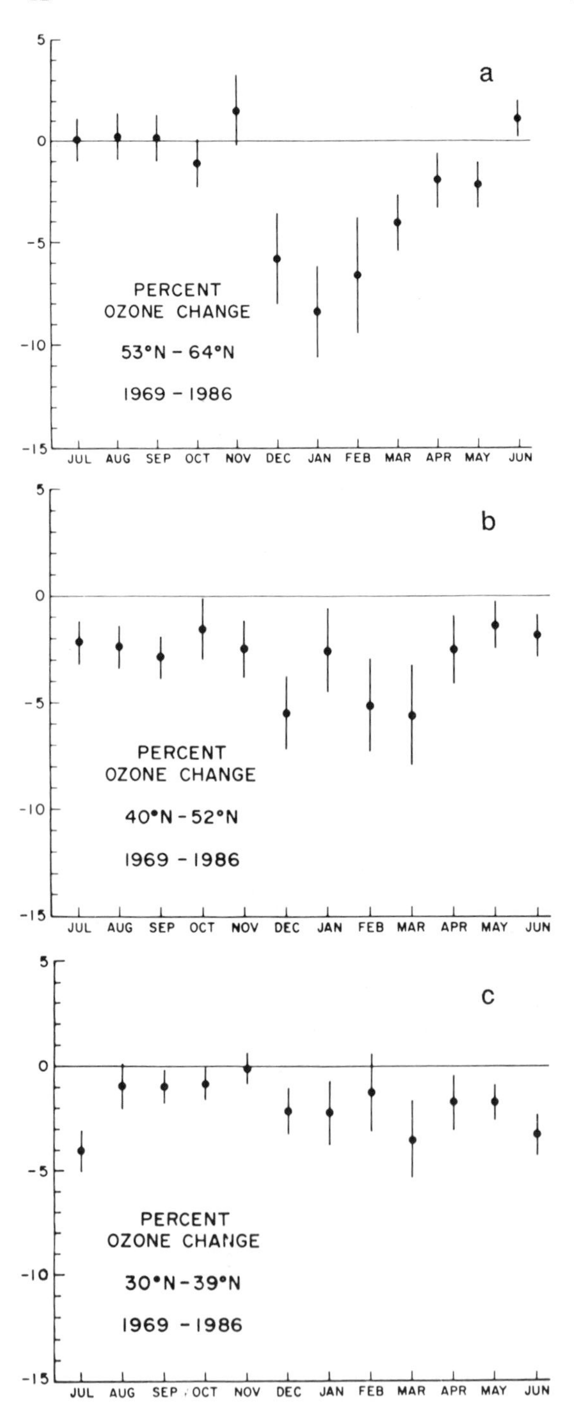

FIGURE 5-7 Percent change in total column ozone between 1969 and 1986 as a function of month for three Northern Hemisphere latitude bands. Estimates of uncertainty are shown by vertical lines. See text for additional explanation.

trend in the regression coefficients appears in several months for the 40°N to 52°N band (Figure 5-8). Detailed examination shows the coefficients tending to be more negative with added years of data, suggesting that the ozone decreases may not be quite linear but may tend toward steeper trends in later years. The great El Chichon volcanic eruption in 1982 does not seem to have had any large effect on the regression coefficients (Dütsch, 1985).

In conclusion, the amount of measured ozone loss in the summer is in reasonable agreement with theory, but the amount of loss poleward of 40°N latitude in winter is greater than that indicated by theoretical calculations. The statistical analysis suggests that there is something missing from the atmospheric models of ozone depletion that affects ozone levels in the Northern Hemisphere in winter. A speculation is that stratospheric ice clouds in the arctic region are having an effect somewhat similar to their effect in the Southern Hemisphere, even though the meteorological conditions are very different in the two hemispheres. The fact that the wintertime ozone decrease in the Northern Hemisphere diminishes gradually at lower latitudes, rather than abruptly as in the Southern Hemisphere, suggests that atmospheric circulation is causing dilution in the Northern Hemisphere.

In the statistical analyses that were reported previously, e.g., the "no statistical change" result in the WMO-NASA report (WMO-NASA, 1986), the assumption was routinely made that the amount of long-term change was independent of month. Therefore, statistical analysts tried to fit all months with a trend having the same slope. Because the statistical reproducibility was much greater in the summer months than in the winter months (see Figure 5-5), the calculations tended to emphasize the summer months in the combined data and led to the conclusion that ozone concentrations were not changing much overall. Our studies show that, when the winter data are analyzed separately, a significant loss of ozone has occurred during the winter months.

(In response to a question about the comparison of Dobson ground instrument data with satellite data): The basic problem with calibrating the satellite data is the known fact that the instrument's diffuser plate has been degrading under bombardment from the sun since launch in October 1978. Calibration is carried out periodically when the satellite passes over a ground-based Dobson instrument site. The ground-based instruments, in turn, are calibrated with the

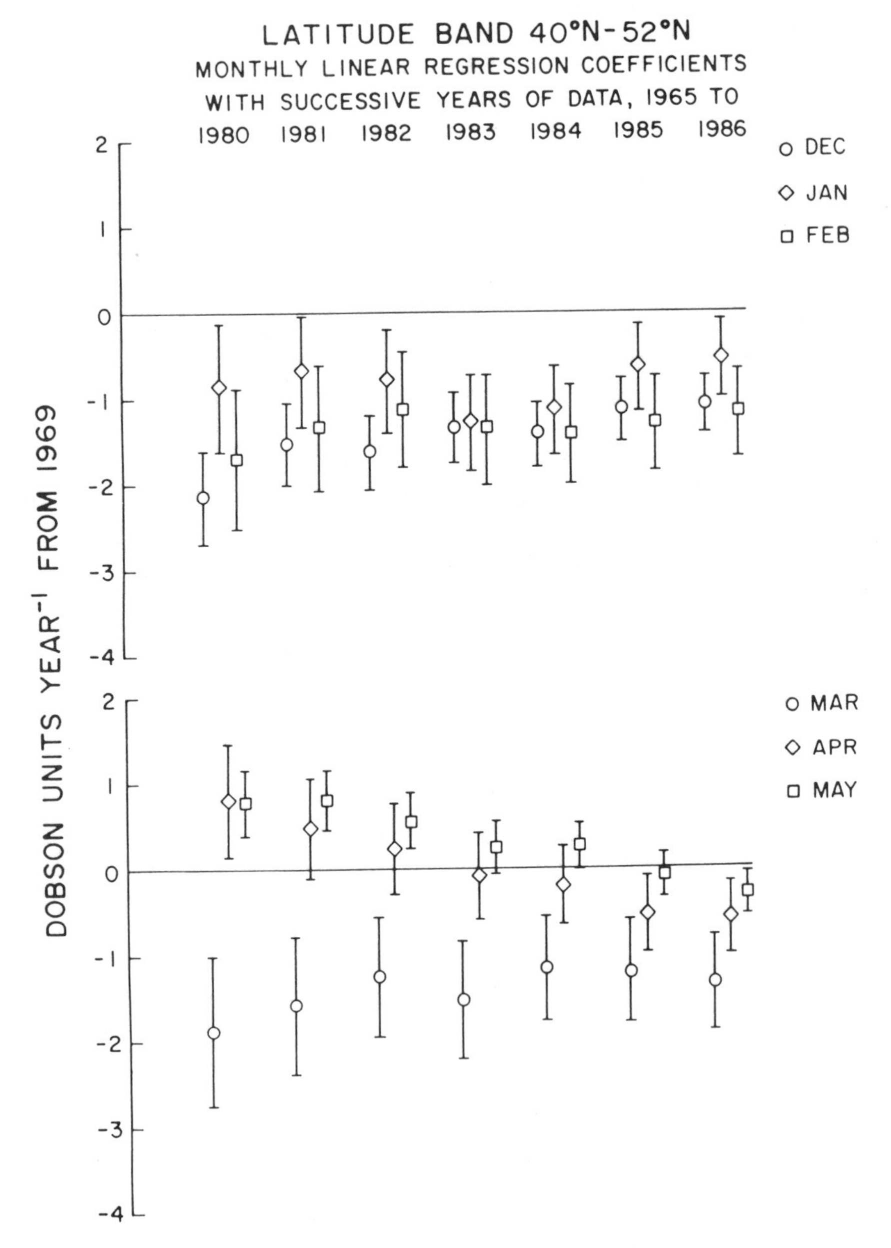

FIGURE 5-8 Computed regression coefficients for linear change in ozone concentrations after 1969, using data for the periods 1965 to 1980, 1965 to 1981, and 1965 to 1982 through to 1986 for each calendar month for the latitude band 40°N to 52°N. Estimates of uncertainty are shown by vertical lines. See text for additional explanation.

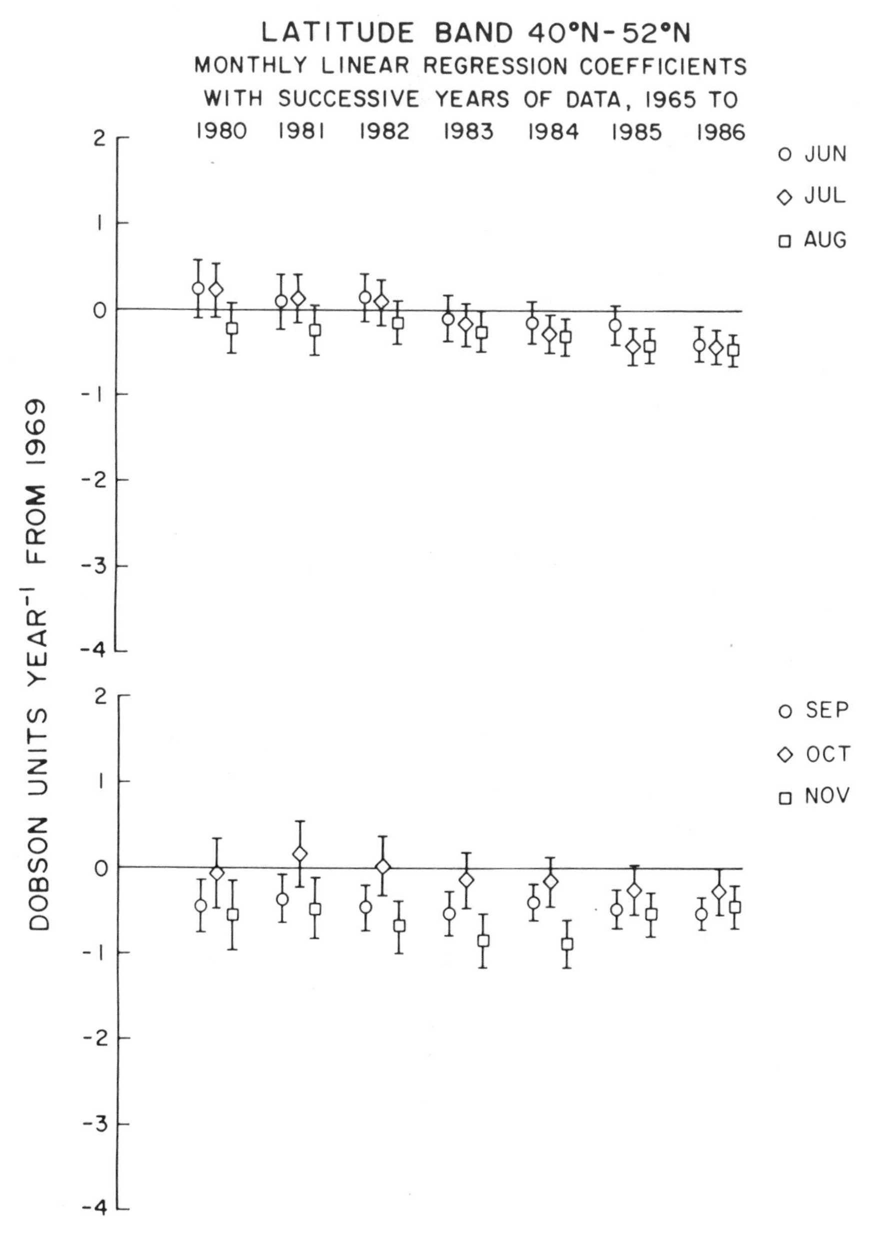

FIGURE 5-8 (continued).

world-standard instrument at Mauna Loa, and the satellite instrument is also compared directly with the Mauna Loa instrument on some overpasses. From these comparisons, it is possible to estimate that the total degradation in the satellite instrument since October 1978 has been about 3.5 percent.

Comparing the average readings from 1979 to 1980 with those from 1986 to 1987 (two years are used to remove the effects of the QBO) shows a loss of ozone in most parts of the world. However, because the solar cycle went from a maximum in the period 1979 to 1980 to a minimum in 1986, a general decline is predicted during this time period. The Dobson data, on the other hand, span 22 years or more, and the effect of the solar cycle can be statistically removed from the data. The solar cycle cannot be reliably removed from the satellite data.

(In answer to a question about the role of tropospheric ozone in the total ozone measurements): The fraction of ozone in the troposphere is approximately 10 percent of the total. There are indications that the amount of tropospheric ozone has been increasing at a rate of about 1 percent per year. Hence, the tropospheric contribution is increasing total ozone at a rate of about 1 percent per decade. If correct, this means that the stratospheric ozone losses are somewhat greater than the total ozone column measurements indicate because of the increase in tropospheric ozone. However, there is some uncertainty, because it is not well known how uniform the amount and rate of increase of tropospheric ozone are globally.

(In answer to a question about the Umkehr ozone data): The Ozone Trends Panel (Watson et al., 1988) has analyzed the Umkehr data and finds a loss of ozone at 40 km of −9 percent, compared to a loss of −3 percent as measured by the Stratospheric Aerosol and Gas Experiment (SAGE) satellite instrument. Theory, depending on latitude and season, predicts a loss of from −5 to −12 percent. So, there is plausible agreement but also concern over whether the Umkehr and SAGE instruments are measuring the same thing.

REFERENCES

Bloomfield, P., G. Oehlert, M.L. Thompson, and S. Zeger. 1983. A frequency domain analysis of trends in Dobson total ozone records. J. Geophys. Res. 88:8512-8522.

Cunnold, D.M., R.G. Prinn, R.A. Rasmussen, P.G. Simmonds, F.N. Alyea, C.A. Cardelino, A.J. Crawford, P.J. Fraser, and R.D. Rosen. 1986. Atmospheric lifetime and annual release estimates for $CFCl_3$ and CF_2Cl_2 from 5 years of ALE data. J. Geophys. Res. 91:10797-10817.

Dütsch, H.U. 1984. An update of the Arosa ozone series to the present using a statistical instrument calibration. Q. J. R. Meteorol. Soc. 110:1079-1096.

Dütsch, H.U. 1985. Total ozone in the light of ozone soundings, the impact of El Chichon. Pp. 263-268 in Atmospheric Ozone (Eds. C.S. Zerefos and E. Ghazi). D. Reidel Co., Dordrecht, The Netherlands.

Rasmussen, R.A., and M.A.K. Khalil. 1986. Atmospheric trace gases: trends and distributions over the last decade. Science 232:1623-1624.

Reinsel, G., G.C. Tiao, M.N. Wang, R. Lewis, and D. Nychka. 1981. Statistical analysis of stratospheric ozone data for the detection of trend. Atmos. Environ. 15:1569-1577.

Reinsel, G., G.C. Tiao, J.L. DeLuisi, C.L. Mateer, A.J. Miller, and J.E. Frederick. 1984. Analysis of upper stratospheric Umkehr ozone profile data for trends and the effects of stratospheric aerosols. J. Geophys. Res. 89:4833-4840.

Rowland, F.S., and M.J. Molina. 1976. Estimated future atmospheric concentrations of CCl_3F (Fluorocarbon-11) for various hypothetical tropospheric removal rates. J. Phys. Chem. 80:2049-2056.

St. John, D., W.H. Bailey, W.H. Fellner, J.M. Minor, and R.D. Sull. 1982. Time series analysis of stratospheric ozone. Commun. Stat., Part A 11:1293-1333.

Watson, R.T., M.J. Prather, and M.J. Kurylo. 1988. Present State of Knowledge of the Upper Atmosphere 1988: An Assessment Report. NASA Reference Publication No. 1208. National Aeronautics and Space Administration, Washington, D.C.

World Meteorological Organization-National Aeronautics and Space Administration (WMO-NASA). 1986. Atmospheric Ozone 1985: Assessment of Our Understanding of the Processes Controlling Its Present Distribution and Change. Global Ozone Research and Monitoring Project, Report No. 16, 3 vols., WMO, Geneva.

6
Heterogeneous Chemical Processes in Ozone Depletion

MARIO J. MOLINA
Jet Propulsion Laboratory
National Aeronautics and Space Administration

This discussion of heterogeneous chemistry focuses first on the reaction between hydrogen chloride (HCl) and chlorine nitrate ($ClONO_2$). By the term "heterogeneous chemistry" is meant a reaction that does not occur only in the gas phase but instead requires a condensed phase in order to proceed. As the previous speakers have indicated, free chlorine atoms are believed to be the principal agent for destroying ozone. Hydrogen chloride and chlorine nitrate species act as sinks, or reservoirs, of chlorine in the atmosphere. These species are not directly reactive with ozone. Hydrogen chloride is a well-known species with well-known properties. Chlorine nitrate is more esoteric, requiring a description of some of its properties.

Chlorine nitrate has two peculiar properties in the context of atmospheric chemistry. One is that in contrast to most other species containing chlorine and composed of many atoms, it absorbs light relatively inefficiently; hence it is an efficient sink for storing chlorine. Related species with fewer oxygen atoms photolyze much more readily and do not serve as efficient reservoirs of chlorine. The other interesting property of chlorine nitrate is that it is a very difficult species to synthesize in the laboratory. On the surfaces of reaction vessels, it is chemically unstable and decomposes rapidly. Thus it came as a surprise to many atmospheric chemists that it was a comparatively stable chemical in the stratosphere. If it reacts with hydrochloric acid on a suitable surface, then the molecular chlorine gas that is produced is quickly broken down into chlorine atoms by the action of absorbed light.

In the context of antarctic chemistry, when this reaction occurs on ice surfaces, the other product of the reaction, nitric acid, remains bound to the ice. The atomic chlorine, on the other hand, reacts with ozone to produce chlorine monoxide and molecular oxygen, destroying two ozone molecules in the process. The chlorine monoxide combines with nitrogen dioxide to reform chlorine nitrate, thereby partially replenishing the chlorine reservoir and removing any remaining nitrogen dioxide from the air. The net effect of the cycle is to release chlorine from the hydrogen chloride (normally very stable), to scavenge nitrogen from the atmosphere, and to generate chlorine monoxide while simultaneously converting ozone to molecular oxygen.

Before the discovery of the antarctic ozone hole, the reaction between hydrogen chloride and chlorine nitrate was considered to be improbable for several reasons. The principal surfaces believed to be effective in the stratosphere in promoting the reaction were those of sulfuric acid, but they appeared to be only weakly favorable for making the reaction proceed. In addition, a simultaneous collision of two gaseous molecules on a surface appeared to be necessary in order for the reaction to occur. Such a simultaneous collision has a very low probability of occurring.

In 1986, we began to study reactions that could be taking place in the polar night stratosphere over Antarctica. We and others surmised that stratospheric ice clouds might be important in causing reactions, since such stratospheric clouds are found almost exclusively in the Antarctic in winter. When it does get cold enough for ice clouds to form, the crystals are not just pure ice. The conclusions of other researchers were that nitric acid would be bound to the solid ice crystal, even at temperatures a few degrees above freezing, through the formation of various hydrates of nitric acid, but that hydrochloric acid would not condense and hence not be bound.

S. Pickering (1893) published a study on the formation of pure ice and of hydrogen chloride hydrate crystals at low temperatures by freezing aqueous solutions. The result of his study was that the concentration by weight of hydrogen chloride compared to water must be greater than about 24 percent for any of it to be incorporated into the frozen crystalline form. This would imply that hydrogen chloride at usual atmospheric concentrations would have no affinity for ice.

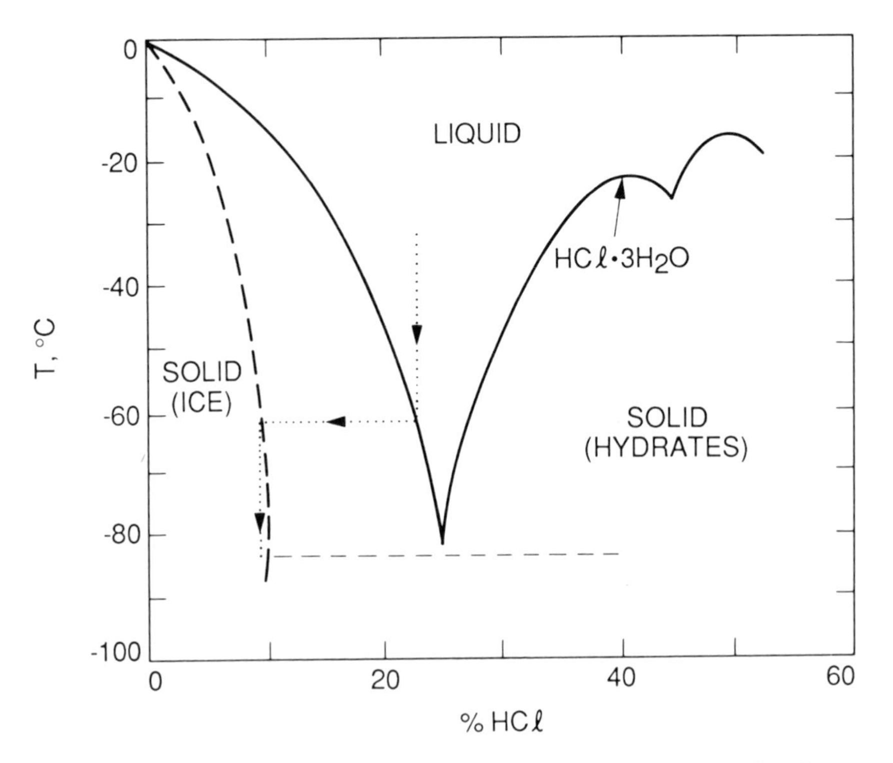

FIGURE 6-1 Equilibrium phase diagram for the hydrogen chloride (HCl) water system as a function of temperature and HCl concentration.

Other studies conducted at temperatures close to freezing came up with the same result.

We carried out experiments to check these results. We measured the hydrogen chloride concentration directly in the solid and in the liquid phase at temperatures appropriate to achieve equilibrium. For initially liquid solutions with a concentration of less than 24 percent, the resultant hydrogen chloride concentration in the solid phase was between about one-third and one-fourth of the corresponding liquid-phase value (Figure 6-1). This result, in contradiction to those of the earlier studies, implies that polar stratospheric ice clouds will absorb significant amounts of hydrogen chloride vapor.

We concluded that ice is actually very efficient in scavenging gas-phase hydrogen chloride. We therefore expect that most of the hydrogen chloride in the antarctic stratosphere is in the condensed phase and bound to ice crystals when they are present. The nature of the nitric acid hydrates that are likely to be present in the antarctic ice crystals needs further study, as the chemistry involved appears to be more complicated than was originally thought. Nevertheless, the end result is simply that both hydrogen chloride and nitric acid are efficiently scavenged by stratospheric ice particles.

We thus realized that if the hydrogen chloride is already in the ice, a simultaneous collision of two species on a surface is not required to cause the hydrogen chloride-chlorine nitrate reaction to occur. Instead, the collision of a chlorine nitrate gas molecule with an ice particle containing hydrogen chloride, an event with a much higher probability of occurring, might be sufficient.

We first carried out experiments to measure diffusion rates for hydrogen chloride in ice at temperatures around 200 K. We used optical absorption techniques to determine the rate of penetration of hydrogen chloride into ice. We were surprised to find that hydrogen chloride moves extremely rapidly within ice. This, we believe, is a result of the relative "openness" of the ice crystalline lattice, combined with the fact that hydrogen chloride molecules are comparatively small, allowing their rapid movement through the lattice. We concluded that hydrogen chloride diffuses almost as quickly in solid ice at 200 K as it does in liquid water, in contrast to the conventional view that diffusion in solids is always much slower. We measured movements of several millimeters on a time scale of just minutes.

The next experiment we did was to look at the infrared spectra of ice samples. We compared the spectrum of pure ice with ice that contained a fraction of a percent of hydrogen chloride: it can be used as a diagnostic to determine if there is hydrogen chloride, even in very small amounts, in ice crystals (Figure 6-2). We did the same experiment with nitric acid. When it is first deposited on ice, the spectrum is very similar to that for condensed nitric acid. After about an hour, the spectrum changes, to show a structure around wave number 700. This spectrum change indicates that the nitric acid is being absorbed by the ice. Finally, the spectrum for ice containing hydrogen chloride and chlorine nitrate is essentially the same as that for ice containing hydrogen chloride and nitric acid (Figure 6-3). This implies that when chlorine nitrate is deposited on an ice crystal that was previously treated with hydrogen chloride, nitric acid is produced that remains in the solid phase within the ice.

We also carried out experiments with ice-coated tubes in order to determine the probability of interaction with ice of several gases. For hydrogen chloride, we determined that at least one of every five

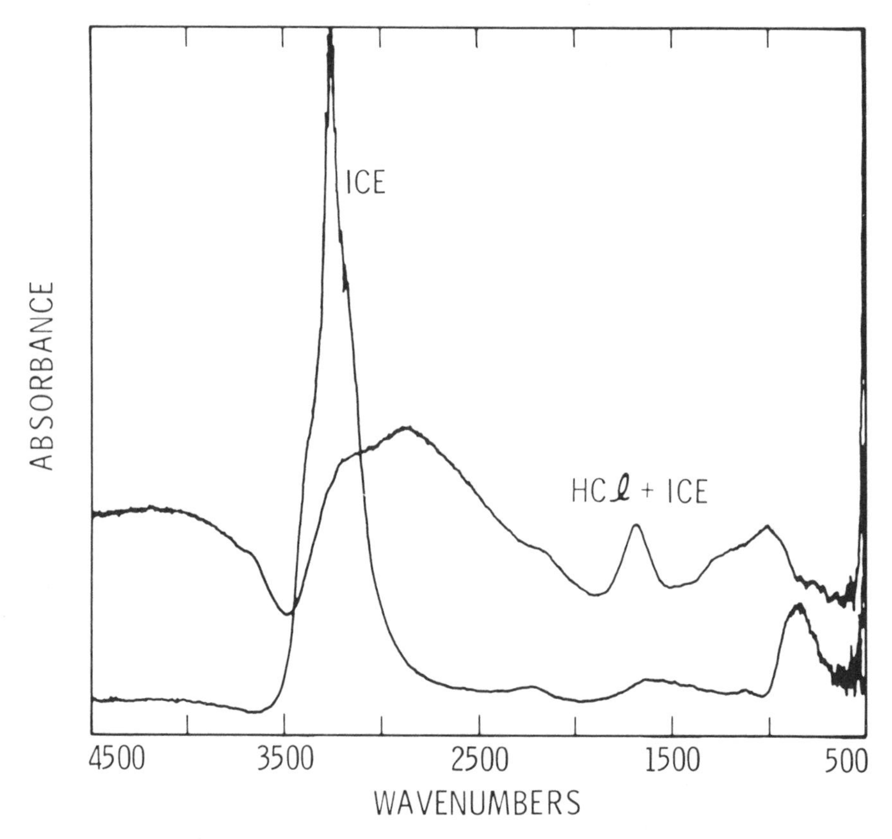

FIGURE 6-2 Infrared spectra of pure ice compared with that for ice containing approximately 1 percent HCl. (Reprinted, by permission, from Molina et al., 1988. Copyright ©1988 by The American Association for the Advancement of Science.)

collisions resulted in the hydrogen chloride being scavenged by the ice. For chlorine nitrate, we also found that the "sticking coefficient" is very large if the ice on the surface of the tube is doped with hydrogen chloride. We also measured molecular chlorine as a product of the collisions. The measured reaction rate is sufficient to explain the observations in the antarctic stratosphere.

The conclusion of all these experiments is that the chlorine nitrate-hydrogen chloride reaction is very efficient in the presence of ice and produces molecular chlorine, thus converting chlorine from an inactive reservoir form to a form that is readily affected by ultraviolet (UV) radiation (Figure 6-4). There are two other reactions that

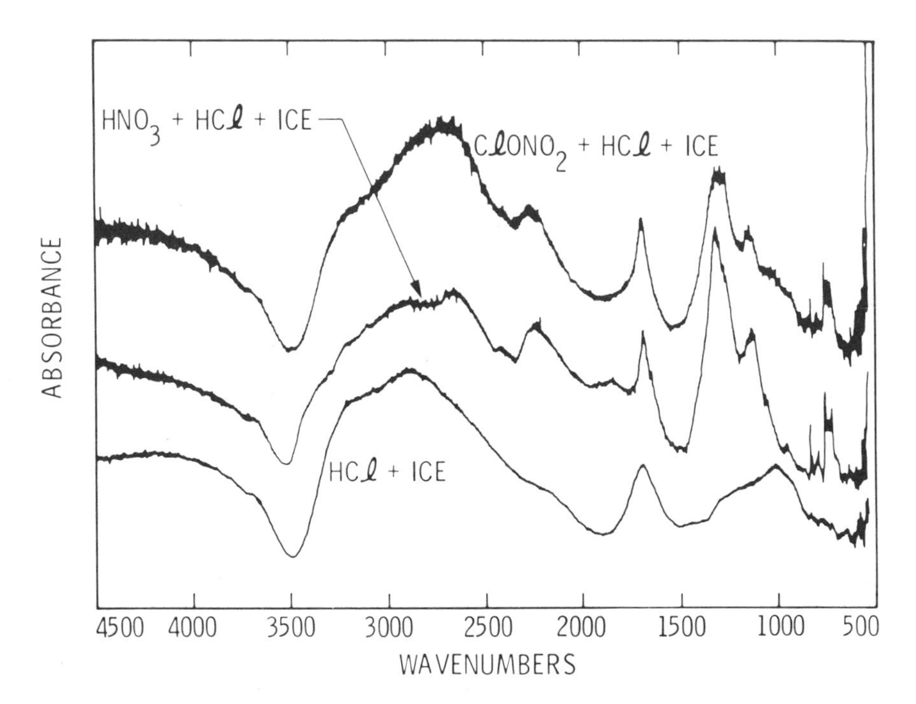

FIGURE 6-3 Comparison of infrared spectra for ice exposed to small amounts of the following compounds: HCl (approximately 1 percent), HCl together with nitric acid (HNO_3), and HCl together with chlorine nitrate ($ClONO_2$). (Reprinted, by permission, from Molina et al., 1988. Copyright ©1988 by The American Association for the Advancement of Science.)

can also result in the "activation" of chlorine. One is the reaction of nitrogen pentoxide (N_2O_5) with the hydrogen chloride in the ice to produce nitric acid and nitryl chloride, which is relatively unstable in the presence of UV. This reaction has been studied by the SRI group and also by a group at the Jet Propulsion Laboratory and has been shown to be an important one for chlorine conversion. The second reaction is that of hypochlorous acid (HOCl) with hydrogen chloride to produce molecular chlorine and water. This reaction also proceeds fairly quickly. Thus, the presence of other species besides chlorine nitrate can result in liberation of the chlorine from the hydrogen chloride in the ice.

Clearly, the presence of ice crystals in the atmosphere is very favorable for the release of chlorine that can destroy ozone. A question that remains is the relative efficiency of ice that contains nitric acid,

$$HCl + ClONO_2 \xrightarrow{ice} HNO_3(s) + Cl_2$$

$$Cl_2 + h\nu \rightarrow 2Cl$$

$$2[Cl + O_3 \rightarrow ClO + O_2]$$

$$ClO + NO_2 \xrightarrow{M} ClONO_2$$

NET:

$$HCl + 2O_3 + NO_2 \xrightarrow{ice} HNO_3(s) + 2O_2 + ClO$$

$$[HCl] > \frac{1}{2}[NO_2]$$

FIGURE 6-4 Series of reactions that result in the destruction of ozone (O_3) molecules by chlorine compounds in the presence of polar stratospheric ice clouds. The net result of these reactions is also shown.

compared to pure ice, in scavenging hydrogen chloride and promoting a reaction with chlorine nitrate. We have done some preliminary studies to answer this question. When we doped the ice with large amounts of nitric acid, the reaction no longer occurred. With progressively smaller amounts of nitric acid, the reaction rate increased. A very similar situation probably obtains with sulfuric acid. The critical point is the amount of water available in the condensed phase for the reaction to occur. Here, we can make a thermodynamic argument: if the temperature is very close to the freezing point of water, even if not to the point where pure ice can be crystallized, the existence of a condensed phase will be sufficient to promote these reactions. Therefore, it seems that any of these acid solutions that is sufficiently close to the frost point of water will promote the reactions in the same way as we have shown.

(In answer to a question on the possible role of large volcanic eruptions in producing sulfuric acid-water ice): The extent to which sulfuric acid droplets would be effective in promoting these types of reactions would depend on the available water that the droplets contain. Concentrated sulfuric acid (95 percent acid) has the small amount of water present so tightly bound that the water is not available for reaction. Fairly dilute solutions may be effective in promoting reactions; laboratory experiments will be needed to determine this. A plausible argument is that if sulfuric acid droplets exist at temperatures close to the frost point of water in the surrounding atmosphere, then they will become very dilute. What happens will depend critically on the temperature, and experiments should be conducted with that in mind.

(In answer to a question about the effects of methane increase and greenhouse gas cooling on stratospheric water-vapor content): Methane and carbon dioxide are usually considered to be the "good guys" in terms of countering the loss of ozone due to chlorine. But the effect of these gases is the opposite for the kind of heterogeneous chemistry that I have discussed. Warming of the surface and lower atmosphere implies cooling of the stratosphere. Both the temperature decrease and the additional water vapor, formed from methane, that reaches the stratosphere will favor greater formation of stratospheric ice clouds.

REFERENCES

Molina, M.J., T.-L. Tso, L.T. Molina, and F.C.-Y. Wang. 1988. Antarctic stratospheric chemistry of chlorine nitrate, hydrogen chloride, and ice: release of active chlorine. Science 238:1253-1257.

Pickering, S. 1893. Die hydrate der chlorwasserstoffsäure. Ber. Dtsch. Chem. Ges. 26:277-289.

7
Free Radicals in the Earth's Atmosphere: Measurement and Interpretation

JAMES G. ANDERSON
Harvard University

This talk deals with the more mechanistic aspects of the ozone depletion problem and includes a discussion of three areas. The first area deals with middle and upper stratosphere homogeneous gas-phase catalysis. I will review some recent results that deal with the fundamental chemical structure of the stratosphere and our understanding of the way these processes occur in the actual atmosphere. The second area covers the topic already discussed by Robert Watson: antarctic ozone depletion. Finally, I will report some results obtained in early 1988 regarding aircraft flights in the Northern Hemisphere that reached 61°N latitude.

Starting with the first topic, I will review our understanding of the gas-phase processes that take place. Molina and Rowland (1974) suggested that the couplet, chlorine atoms plus ozone reacting to form chlorine monoxide (ClO) plus oxygen followed by chlorine monoxide reacting with oxygen atoms to reform chlorine atoms plus oxygen in a catalytic cycle, constitutes a loss process for ozone in the stratosphere. The rate-limiting step in that couplet is the slower of the two reactions that controls the frequency of closure of the catalytic cycle. A definition of the distribution of chlorine monoxide at midlatitudes is now complete between the tropopause and the stratopause. Combined measurements from aircraft and balloon-borne instruments show that the top of the profile varies, we believe in response to methane. This brings up the very crucial coupling between the catalytic cycle and the ability of the atmosphere to deliver molecules from the tropopause region to the upper stratosphere. If circulation slows down and methane is not delivered to the upper stratosphere, perhaps because of distorted heating patterns, the efficiency of these mechanisms will change dramatically.

In the lower stratosphere, on the other hand, we find through observed diurnal variations of the chlorine monoxide radical that there is coupling between the nitrogen and chlorine families. As Daniel Albritton has previously indicated, the presence of chlorine in the atmosphere extracts radicals of the nitrogen family from the atmosphere. In the normal atmosphere, unpolluted by chlorine radicals, nitrogen oxides account for some 70 to 75 percent of globally integrated ozone destruction. Thus in the midlatitudes there has been somewhat of a canceling effect; this has had an impact on the history of ozone studies. To properly treat the entire stratospheric ozone system requires consideration of some 200 gas-phase reactions coupled with transport processes. If we can observe free radicals in situ, we can simplify the problem considerably. We can talk about the effect of increases in chlorofluorocarbons (CFCs) continuing into the next century. If other factors are held constant, the mixing ratio of the sum of atomic chlorine and chlorine monoxide will be raised as the CFCs reach the stratosphere, thereby raising the efficiency of the rate of chlorine radical catalysis. The coupling with the nitrogen system occurs when the chlorine monoxide plus nitrogen dioxide (NO_2) reaction occurs, forming chlorine nitrate ($ClONO_2$). The partitioning between hydrogen chloride and chlorine radicals is controlled in a positive sense by the hydroxyl radical (OH). The more OH that exists in the upper atmosphere, the higher will be the fractional amount of total chlorine tied up in the chlorine radical form. The reverse is true in the nitrogen system, in which OH reacting with nitrogen dioxide forms nitric acid. The OH concentration in a highly simplified form is established by a balance between excited oxygen atoms, $O(^1D)$, reacting with water and (at 30 km and below) the catalytic conversion of OH back to water, with nitric acid as the catalytic agent.

So we can, in turn, examine the processes that we believe will dominate into the next century by starting with increased CFCs. The CFCs cause chlorine monoxide to increase, and the chlorine monoxide in turn controls the nitrogen dioxide concentration through the ClO plus NO_2 reaction. The nitric acid concentration suffers as a result. Reductions in the nitric acid concentration drive up the OH concentration, leading to a positive feedback. The feedback is established first through the OH plus HCl reaction, and second through the diminished nitric oxide (NO) concentration, which blocks the channel from chlorine monoxide back to chlorine.

As we look into the future, we have to understand the coupling between chlorine and nitrogen and also the processes that control the OH concentration. In February 1988, we flew a series of aircraft experiments with a chlorine monoxide instrument on board (see Figure 7-1) to map out the diurnal behavior of chlorine monoxide at midlatitudes over California. The detection threshold of the instrument is such that chlorine monoxide at 1 part per trillion is easily detectable. We followed the build up of chlorine monoxide at sunrise over 2 orders of magnitude. We obtained similar measurements in the evening, allowing us to identify the coupling reactions. The first analysis with a model shows that, in fact, the dominant process involved chlorine nitrate, from the reaction of nitrogen dioxide with chlorine monoxide, and the photolysis of chlorine nitrate back into free radical form. That tells us that the chlorine and nitrogen families are indeed coupled, a fact that is crucial as the system is loaded with increasing amounts of CFCs.

We turn to the next question, that of the OH radicals, and history is again an important teacher. The sequence of integrated column ozone depletion predictions, which began in 1976, traces a clear reduction in column concentration of ozone after the introduction of chlorine reactions in the prediction models. After 1978, when agreement on the reactions used in models was reached, we see a profound effect on the model results of increasing the reaction rate constant of $NO + HO_2 \rightarrow NO_2 + OH$, which increased the model-calculated column integral of ozone depletion by more than a factor of three. Subsequent refinements, such as the introduction of OH loss terms (nitric acid plus OH and pernitric acid plus OH), had a further impact on predicted column ozone depletion. The model predictions can be checked by observing the concentration of OH in the lower stratosphere.

To carry out that check, we flew an experiment during the summer of 1987 on a high-altitude (to 42.7 km) research balloon, which used a high-repetition-rate copper vapor laser system that is controlled by command from the ground through computer algorithms. These allow tuning in a controlled way on a balloon platform, thus permitting laboratory-quality experiments from these platforms. For example, we can superpose the OH spectrum from a plasma discharge lamp with the output from the laser, defining precisely which

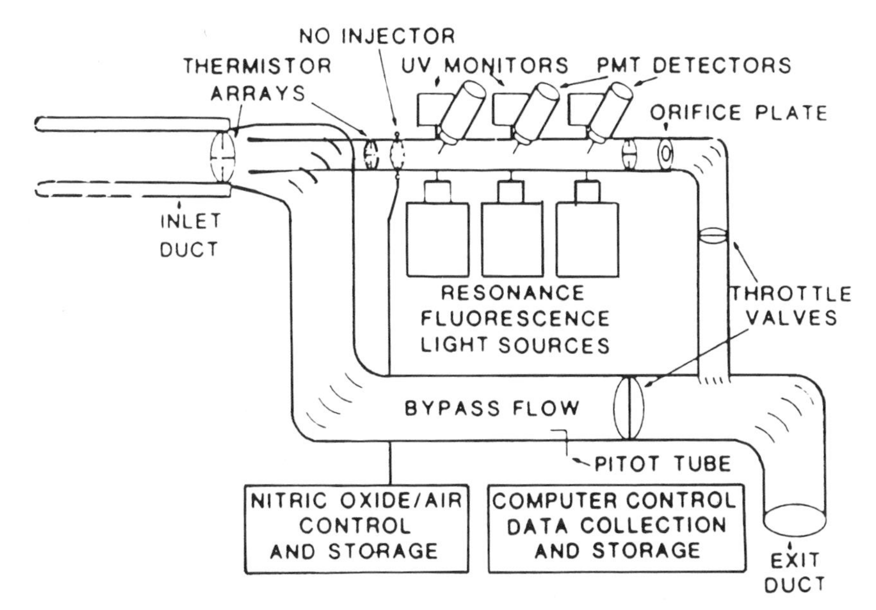

FIGURE 7-1 Schematic of the instrument developed for the ER-2 aircraft to observe chlorine monoxide (ClO) and bromine monoxide (BrO) in situ within the antarctic polar vortex. (Adapted from Brune et al., 1989.)

rovibronic transition is being pumped during the atmospheric measurement. This is simply one example of a number of cross-checks that allow us to believe what we are observing. The outcome of the measurement is as follows: if we look at the mixing ratio of OH as a function of altitude in the crucial region between 22 and 30 km, we can use a model by Ko, Tung, Weisenstein, and Sze (1985) that includes nitric and pernitric acid. Figure 7-2 shows that observed OH agrees closely with the model-calculated distribution in the lower stratosphere. This implies that under present-day conditions, one would expect less change in ozone for a given amount of chlorine. But that is not true because OH, we now believe, is controlled by nitric and pernitric acid, which means it is susceptible to linkage through the chlorine-induced removal of nitrogen oxides. So, while the OH concentration is low, it is susceptible to change with increasing mixing ratios of total chlorine in the atmosphere. In conclusion, many of the basic elements of our picture of midlatitude catalytic destruction of ozone are now coming into focus.

In the meantime, we have the antarctic ozone hole, which now

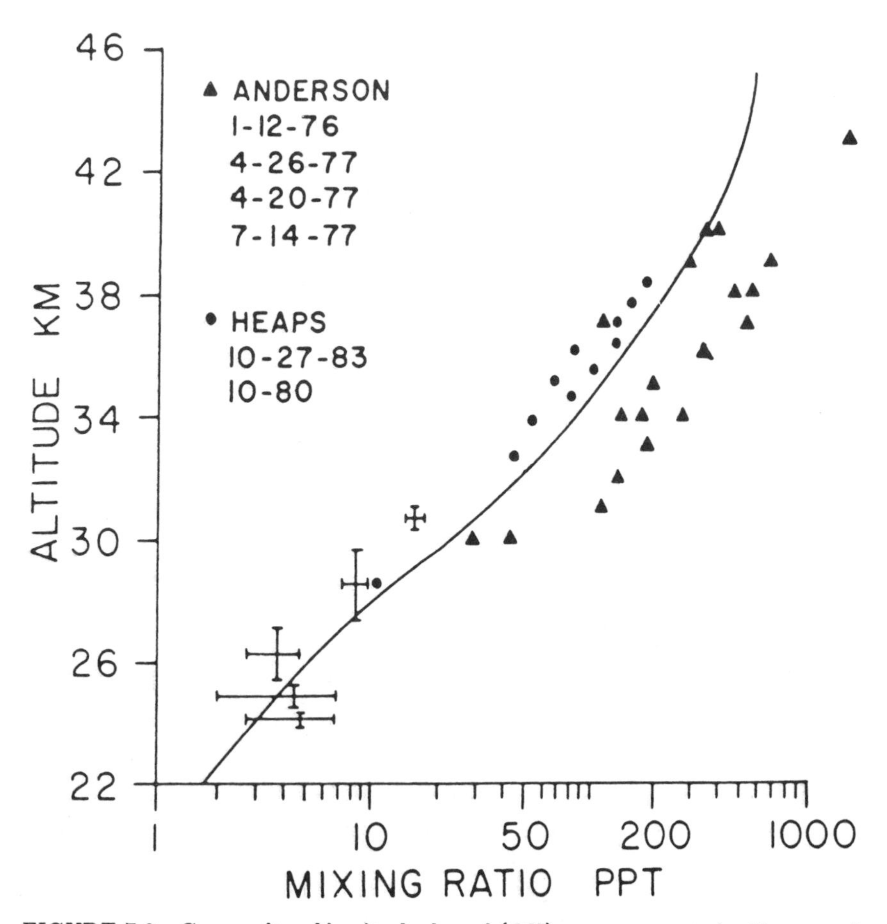

FIGURE 7-2 Composite of in situ hydroxyl (OH) measurements by Heaps and McGee (1985), Anderson (1976), and this work (with approximate error bars shown). The calculation (Ko et al., 1985) shown by the solid curve corresponds to simulation of the present-day atmosphere for July conditions at 30°N. Water vapor is fixed in the model from 4.5 ppmv in the lower stratosphere to 6.0 ppmv in the upper stratosphere.

is receiving significantly more focused attention. As Watson mentioned, their high-altitude ER-2 aircraft flew from the southern tip of Chile (Puntas Arenas), ascended rapidly to 65,000 to 67,000 feet (about 20 km), and then flew approximately horizontally to 72°S latitude. There, the aircraft descended to 45,000 feet (13.7 km) and then climbed back up and returned to Punta Arenas. The first penetration—that descent on August 23, 1987—into the polar vortex showed the chlorine monoxide concentration there to be increased by a factor of 100 over the concentration measured above Puntas Arenas and considered to be representative of midlatitude concentrations at that time. The threshold of the instrument was 0.1 part per trillion, and the response time was about 0.25 s; therefore, our confidence in the measured profiles is very high. At 700 parts per trillion, the measured amount of stratospheric chlorine monoxide corresponds to approximately 200 times normal levels. (In the laboratory, where we study reactions of these chlorine monoxide molecules, we use a concentration of only about 5 parts per trillion to look at the kinetics of the species that exist over Antarctica.)

The question is, what happened to the ozone over Antarctica? On August 23, 1987, the ozone concentration was still unperturbed. Whatever took place on the stratospheric ice crystal structures during the dark winter months had little effect on the ozone concentration. But after August 23, the sun penetrated the South Polar vortex region for a small but increasing number of hours per day. Three weeks later, on September 16, the aircraft profile through the boundary of the polar vortex still showed a dramatic increase in chlorine monoxide on the poleward side. Conversely, the ozone level had dropped markedly on the poleward side. Thus, there was a strong anticorrelation between chlorine monoxide and ozone levels that had developed during the 3-week period. These results are summarized in Figure 7-3. The anticorrelation, by itself, does not prove anything, but the change from an initial condition in which ozone was unaffected by chlorine monoxide on August 23 to the strong negative correlation on September 16 does indicate what kind of process must be occurring. Actually, the behavior of the geochemical system within the polar vortex is about as simple as any likely to be encountered in the atmosphere.

Given 500 times the normal free radical concentration and the strong anticorrelation between ozone and chlorine monoxide that developed in 3 weeks, what are the fundamental mechanisms? There are three possibilities. The one I will discuss in detail involves the formation of a chlorine monoxide dimer. There are two forms: one is a peroxide structure with chlorine atoms at a dihedral angle of 90°; the other is a ClO-ClO structure. Molina and Molina (1987) have suggested the symmetrical (peroxy) dimer as the key structure in the mechanism that may be important in the antarctic region. It replaces the O plus ClO rate-limiting step with a pressure-dependent dimerization step that is followed by photodissociation of that dimer

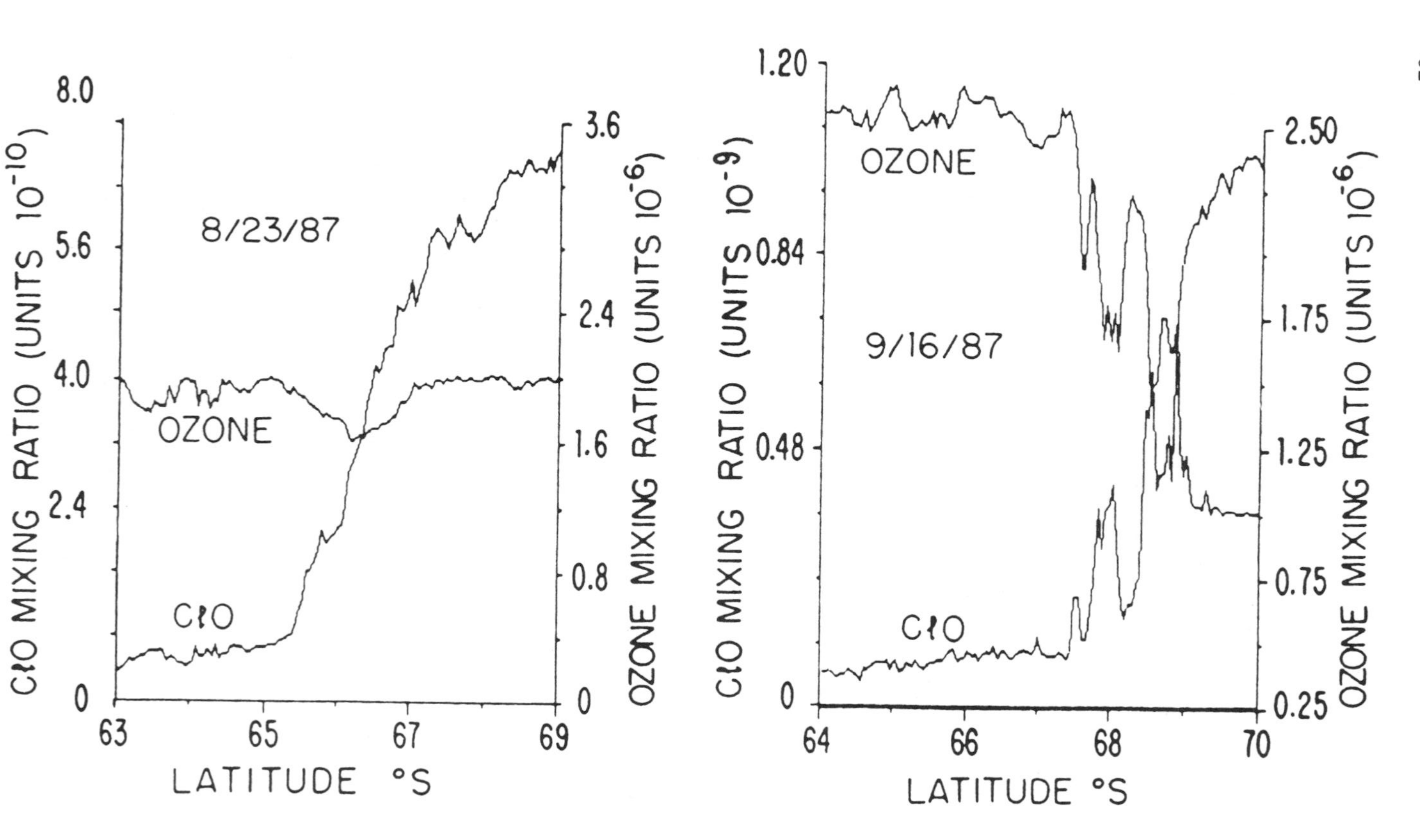

FIGURE 7-3 Evolution of the relationship between ozone and chlorine monoxide (ClO) from the first full penetration into the vortex on August 23, 1987 (left), to three weeks later on September 16, 1987 (right) (Anderson et al., 1989a).

to form Cl + ClOO. The other suggestions revolve around chlorine and bromine and around chlorine and hydrogen.

We can analyze the effect of just the ClO + ClO → ClOOCl step by simply looking at the continuity equation for ozone. There are two chemical contributions and a dynamical contribution. It is possible to set upper limits on the flux divergence terms for ozone within the vortex. We can tell that approximately 5 million molecules of ozone are destroyed per cubic centimeter per second by chlorine monoxide at midday. We can then calculate, in a very simple way, the expected decrease in ozone because of the inherent symmetry of the problem. We have air moving in a circumpolar pattern that deviates somewhat in latitude, but the time constant of that deviation is small compared to the time constant of ozone changes. So, we can take slices into the rotating polar vortex "disk" and sample the chlorine monoxide and ozone concentrations simultaneously. The kinetics are determined by calculating the time rate of change of ozone based on the observed chlorine monoxide concentrations. We correct for the shape of the ClO radical diurnal concentration variation and integrate over each 24-hour period, normalizing to the observed ClO concentration.

We have mapped out the temporal dependence of chlorine monoxide on several potential temperature (isentropic) surfaces as a function of latitude. (Potential temperature surfaces were used to minimize the effects of atmospheric fluctuations.) We can calculate the incremental ozone that is removed each day, based on the in situ observation of ClO. We can take the rate constants for the rate-limiting step in the dimerization cycle of ClO-ClO, put in the observed ClO concentration, and compare the rate of that catalytic cycle with the observed removal of ozone. This procedure may not have the precision of a laboratory experiment, but it is very indicative of the behavior of the system. In fact, for a continental-scale chemical problem, the results are very straightforward. Figure 7-4 summarizes the results of this comparison.

I will pass quickly over the bromine contribution mentioned earlier by Watson. Bromine is present in the vortex at concentrations of about 5 parts per trillion. That corresponds to some 5 to 10 percent of the removal process due to chlorine. Chlorine was clearly the important removal agent in 1987 and will remain so for many years.

I will finish by discussing our latest results for the Northern Hemisphere. Again, Brune was the major collaborator here. If we look at the wintertime latitude-versus-altitude dependence of chlorine monoxide as far north as 61°N latitude and as high as 20 km,

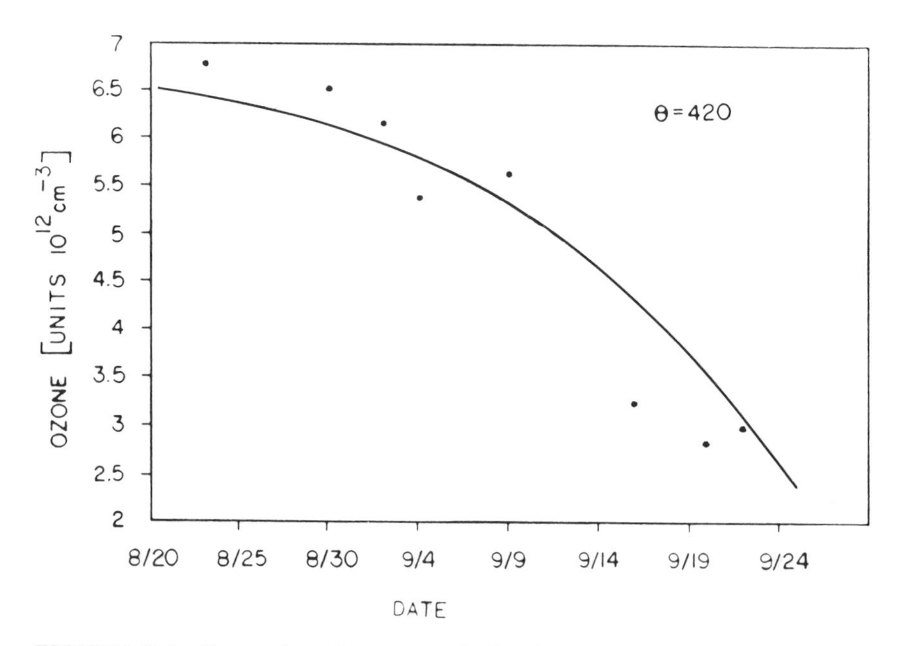

FIGURE 7-4 Comparison between calculated and observed rates of ozone loss at a fixed potential temperature (which eliminates adiabatic variations) throughout the course of the Airborne Antarctic Ozone Experiment (Anderson et al., 1989b).

we note that chlorine monoxide is strongly perturbed from normal conditions, with 60 parts per trillion at the edge of the polar stratospheric jet. Even at 32 to 35°N latitude, the amount is significantly higher than summer levels. In fact, summer levels are generally exceeded to some extent poleward of 25°N latitude. If one compares the recent winter data with mean July data, the difference at 36°N latitude is about a factor of two. The 60 parts per trillion at 61°N is not nearly as high a concentration as the 1,000 parts per trillion observed in the Antarctic, but then the Northern Hemisphere measurements probably did not extend all the way into the polar vortex.

(In answer to a question): I think that the lower temperatures in the absence of sunlight in the Arctic are tying up the oxygenated nitrogen in the form of either nitric acid or nitrogen pentoxide (N_2O_5) and leaving a large deficiency of nitric oxide and nitrogen dioxide. A deficiency of nitrogen dioxide has been observed using ground-based measurements in central Canada. This is probably a manifestation of the gas-phase process that allows chlorine monoxide to increase in the absence of nitrogen dioxide. The degree to which the nitrogen system is perturbed depends on the stability of the polar vortex, which is clearly greater in the Southern Hemisphere than in the Northern Hemisphere.

REFERENCES

Anderson, J.G. 1976. The absolute concentration of OH in the earth's stratosphere. Geophys. Res. Lett. 3:165-168.

Anderson, J.G., W.H. Brune, and M.J. Proffitt. 1989. Ozone destruction by chlorine radicals in the antarctic vortex: The spatial and temporal evolution of ClO - O_3 anticorrelation based on in situ ER-2 data. J. Geophys. Res., Special Issue on Antarctic Ozone (in press).

Anderson, J.G., W.H. Brune, S.A. Lloyd, W.L. Starr, M. Loewenstein, and J.R. Podolske. 1989. Kinetics of O_3 destruction by ClO and BrO within the antarctic vortex: An analysis based on in situ ER-2 data. J. Geophys. Res., Special Issue on Antarctic Ozone (in press).

Brune, W.H., J.G. Anderson, and K.R. Chan. 1989. In situ observations of ClO in the Antarctic: ER-2 aircraft results from 54°S to 72°S latitude. J. Geophys. Res., Special Issue on Antarctic Ozone (in press).

Heaps, W.S., and T.J. McGee. 1985. Progress in stratospheric hydroxyl measurement by balloon-borne LIDAR. J. Geophys. Res. 90:7913-7921.

Ko, M.K.W., K.K. Tung, D.K. Weisenstein, and N.D. Sze. 1985. A zonal mean model of stratospheric tracer transport in isentropic coordinates: numerical simulations for nitrous oxide and nitric acid. J. Geophys. Res. 90:2313-2329.

Molina, L.T., and M.J. Molina. 1987. Production of chlorine oxide (Cl_2O_2) from the self-reaction of the chlorine oxide (ClO) radical. J. Phys. Chem. 91:433.

Molina, M., and F.S. Rowland. 1974. Stratospheric sink for chlorofluoromethanes: chlorine atom catalysed destruction of ozone. Nature 249:810-812.

8
Theoretical Projections of Stratospheric Change Due to Increasing Greenhouse Gases and Changing Ozone Concentrations

JERRY D. MAHLMAN
Geophysical Fluid Dynamics Laboratory
National Oceanic and Atmospheric Administration

This talk discusses what has happened in the stratosphere and what may happen there in the future. I will first review the ensemble of gases present in the stratosphere and their effects:

1. Chlorofluorocarbon (CFC-11 and CFC-12) increase. The increase of CFCs already appears to be causing ozone loss through the action of chlorine. CFCs also act as greenhouse gases in the troposphere.
2. Methane (CH_4) increase. We have heard in a previous presentation that methane is a tropospheric greenhouse gas. The approximately 1 percent per year increase in methane also has implications for long-term increases of the water vapor amount in the middle and upper stratosphere. However, the methane chemistry opposes the chlorine catalysis chemistry in the lower stratosphere. Methane increases also play a role in reducing the amount of hydroxyl (OH) in the troposphere.
3. Nitrous oxide (N_2O) increase. Nitrous oxide is increasing by what appears to be the comparatively modest amount of about 0.2 percent per year. However, the N_2O anthropogenic source is about one-third of the natural source and thus is not negligible on long time scales. Nitrous oxide is a tropospheric greenhouse gas, but its small annual increases are contributing little to current increases in infrared radiative forcing. Reaction of N_2O with excited atomic

oxygen in the middle stratosphere produces reactive nitrogen (NO_x), which provides the major natural catalytic loss of ozone. Oddly, NO_x provides an important negative feedback against the growing attack on ozone by reactive chlorine (Cl_x). This occurs through the formation of chlorine nitrate ($ClONO_2$), thus inhibiting both Cl_x and NO_x catalytic ozone destruction cycles.

4. Stratospheric carbon dioxide (CO_2) increase. The increases in CO_2 will lead to a strong cooling trend in the stratosphere. To some extent, this cooling effect acts to moderate the expected mid-stratospheric ozone decreases.

5. Stratospheric ozone decrease. Large ozone decreases will also result in large stratospheric cooling. There is also strong column "self-healing" of ozone, which I will discuss later.

6. Stratospheric water vapor increase. An increase would result in increased downward infrared radiative flux, complex chemical changes if stratospheric ice clouds form, and an increased ozone loss in the 30- to 50-km layer.

I will next discuss what the NASA-WMO Ozone Trends Panel (Watson et al., 1988) has learned about recent (1979 to 1985) ozone trends in the stratosphere. SAGE satellite data suggest a decrease of about 3 percent near 40 km altitude, on an average worldwide, over the 6-year period. The ground-based Umkehr data, on the other hand, suggest a decrease of about 9 percent at 40 km. Model calculations predict a 4 to 9 percent decrease in response to increased trace gases, primarily CFMs, and a 1 to 3 percent decrease in response to declining solar activity, for a total decrease of 5 to 12 percent. Given this range of uncertainty, the observed ozone changes near 40 km are not inconsistent with theory. The credibility of the theoretical results is enhanced by the observation that stratospheric temperatures have decreased globally, between 25 and 55 km altitude, by 1.7°C since 1979. This decrease is consistent with decreases in upper stratospheric ozone of up to (but not larger than) 10 percent. The vertical profile of ozone change also is in fair agreement with theoretical predictions.

I will now turn to a discussion of future trends. *Atmospheric Ozone 1985* (WMO-NASA, 1986) contains estimates for equilibrium (infinite elapsed time) changes using one-dimensional chemical models and assuming that CFM emissions are held constant at the 1980 rate. At 10 ppb reactive chlorine species, equilibrium total column loss is estimated at between 5 and 9 percent. At 15 ppb chlorine species, the estimated loss is 10 to 20 percent. Thus the equilibrium column loss doubles, for a 50 percent increase in odd chlorine. This nonlinearity results from the progressively greater scavenging of nitrogen oxides by chlorine, leaving the excess chlorine to destroy increasingly more ozone. On the other hand, if carbon dioxide is doubled (with no other atmospheric changes), equilibrium total column ozone will increase by 2 to 3 percent.

However, the changes at the 40-km level predicted by the same model are much more drastic. At a concentration of 10 ppb chlorine species, the equilibrium prediction is an ozone decrease of 60 to 80 percent, and at a concentration of 15 ppb chlorine species, a loss of 70 to 85 percent would result. For the global-mean vertical profile of ozone mixing ratio, the equilibrium prediction is for an *increase* in ozone in the lower stratosphere, between 10 and 25 km, of a few percent, and a *decrease* at higher altitudes, with the maximum decrease at about 40 km. (The feedback of higher temperatures in the lower stratosphere and lower temperatures in the higher stratosphere would reduce slightly the magnitudes of both the increase and decrease.) The predicted increase in ozone below about 25 km would result from more ultraviolet (UV) radiation penetrating to lower altitudes and creating more ozone in a kind of negative feedback process that tends to limit the depletion of total column ozone. Thus the change in the total column ozone is a comparatively small difference between two large numbers, given that the atmospheric mass drops off nearly exponentially with increased altitude.

Figure 8-1 shows the two-dimensional model prediction of percent ozone decrease by Atmospheric and Environmental Research, Inc. (AER) as reported in WMO-NASA (1986) at 8.2 ppb Cl_x species. This is about equal to the equilibrium Cl_x for 1980 emission rates. The resulting stratospheric distribution shows maximum ozone losses at about 40 to 45 km poleward of about 50°N and 50°S latitudes. In these regions, the predicted ozone loss is greater than 50 percent. On the other hand, ozone is predicted to *increase* by about 20 percent in the lower stratosphere (15 to 20 km) near the equator. In low latitudes, the total column self-healing effect is comparatively strong because the incident solar UV is strong all year, whereas the effect is much weaker at the high latitudes because of weaker UV and the presence of large downward mean advection of ozone at these latitudes.

A question that is fair to ask is, what is the credibility of such two-dimensional models? This depends, of course, on which effects

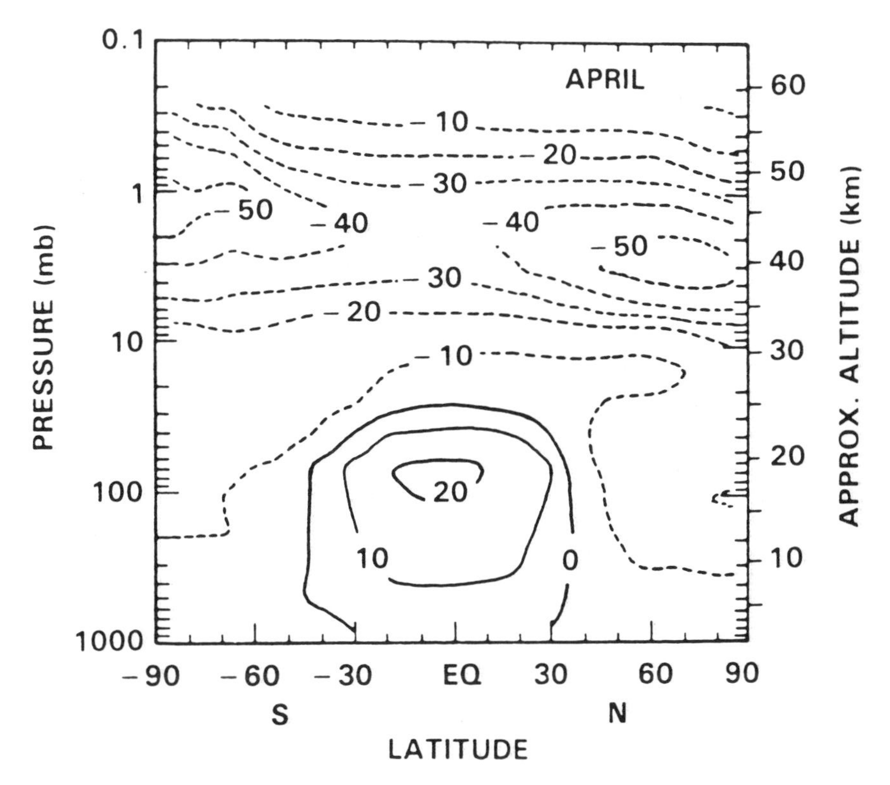

FIGURE 8-1 Two-dimensional model prediction of the percent change in ozone for 8.2 ppbv of total reactive chlorine (Cl_x). Model results from the chemical-transport model of Atmospheric and Environmental Research, Inc. (Reprinted from WMO, 1986.)

are included and which are left out. The models predict large changes in ozone concentrations; therefore, one would expect significant temperature and circulation changes to occur as well. It should be pointed out that two-dimensional models used for ozone assessment all make a similar assumption about change in circulation. They implicitly assume that the stratospheric circulation *does not change* as absorbers are changed. This corresponds to the "fixed dynamical heating" limit; that is, the net heating of local air by dynamical processes *does not change* with changing absorbers, even though the temperature itself is free to change. However, when a real climate system such as the stratosphere is perturbed, the changed distribution of absorbers may or may not lead to a changed temperature distribution in order to maintain what we call "climate balance." The sources of temperature change are not just radiative but may also be dynamical. Dynamical heating may result from advection, diffusion, and adiabatic compression or expansion. The radiative heating is due to short-wave solar radiation (almost independent of atmospheric temperature) and long-wave radiation (strongly a function of temperature). The assumption is routinely made that the stratosphere is in radiative equilibrium, resulting in a very simple system. But we know that this is a gross oversimplification, especially in the higher latitudes, where dynamic effects are very important. Actual temperatures in the polar latitudes sometimes differ by as much as 50 to 60 K from those given by radiative equilibrium models.

Thus if the distribution of absorbers is changed, the dynamical heating, as well as the radiative heating, is likely to change. If dynamical heating is important, three-dimensional models that include dynamical response should be used in preference to the two-dimensional models that implicitly invoke the fixed dynamical heating assumption. Such a two-dimensional model uses a latitude-altitude framework that has mean mass circulation directed upward in low latitudes, downward in high latitudes, and from equator to pole at stratospheric altitudes. Superimposed on the mean meridional motion is a meridional eddy diffusion of particles that is driven by upward-propagating tropospheric disturbances. Both of these transport mechanisms arise in the troposphere; thus the stratospheric latitude-altitude motions (and the degree of departure from radiative equilibrium) are driven by the troposphere and would decay to near zero in the absence of dynamical forcing from the troposphere.

The question now is, how does the stratospheric circulation respond when the distribution of absorbers is changed? Calculations of this type were done at the Geophysical Fluid Dynamics Laboratory (GFDL) in 1980 by Fels et al. In one numerical experiment, carbon dioxide was doubled but sea temperatures were held fixed. In this case, Figure 8-2 shows that the three-dimensional dynamic equilibrium model results agree fairly closely, to within 1°C, with those from a fixed dynamical heating radiative model. The dynamic model predicts cooling of 11°C at 45 km, compared to the radiative model's prediction of about 10°C. Both models show a cooling of about 2 to 4°C between 20 and 25 km. Figure 8-3 shows another experiment, in which an arbitrary uniform reduction of ozone by 50 percent was assumed. (This experiment ignored the column self-healing effect.) Again, the radiative model agrees fairly closely with the dynamic model in terms of magnitude of cooling, but some of the gradient details are radically different, particularly in the tropics. Cooling of 6 to 8°C is predicted in the high latitudes between 15 and 45 km by both models. At the equatorial tropopause, a 12°C cooling is predicted. If we more realistically assume that the ozone at the equatorial tropopause increases as in the AER model, then we may infer that a warming of about 4°C will tend to occur there.

Recently, we have learned much more about how poor a job the older models did of simulating the dynamics of the stratosphere. As the horizontal resolution of models has increased, predicted stratospheric temperatures have increased and come into better agreement with observations. Predicted temperatures using a 1°-latitude resolution dynamic model agree closely with temperatures observed at 62°N latitude in December, whereas results for a 9° resolution dynamic model were fairly close to those obtained with the radiative model but from 15 to 45°C colder than either the observed temperatures or results for the high-resolution dynamic model (see Mahlman and Umscheid, 1987). Thus, the traditional higher-latitude stratospheric cold bias of general circulation models was not the fault of radiative transfer but rather the fault of oversimplifying the dynamics in the models. At high resolution, the models are considerably more dynamical in the stratosphere and produce temperatures and motions that look much more like those in the real stratosphere. It appears that the tropospheric dynamical processes are strong enough to push the stratosphere some distance away from its radiative equilibrium condition.

Thus dynamic modeling is essential for predicting changes correctly in the stratosphere. The good news is that a model with sufficiently high resolution is capable of useful predictions. The bad news is that such models, if run to equilibrium conditions, require substantial computer resources. Possibly, some of the resolution can be traded for carefully devised parameterizations that are self-consistent and appropriately sensitive to climate changes. In a stratosphere that is dynamically driven, the interannual variability is quite large, thereby increasing the overall computational problem.

Based on our experience thus far, I would like to speculate on what I consider to be the stratosphere climate issues that we will have to face beyond what we already know. We think we know that the upper stratosphere will cool by 20 to 25°C, perhaps more; this makes the stratosphere a candidate for inclusion in a full climate system model. Also, there will presumably be large ozone decreases in the

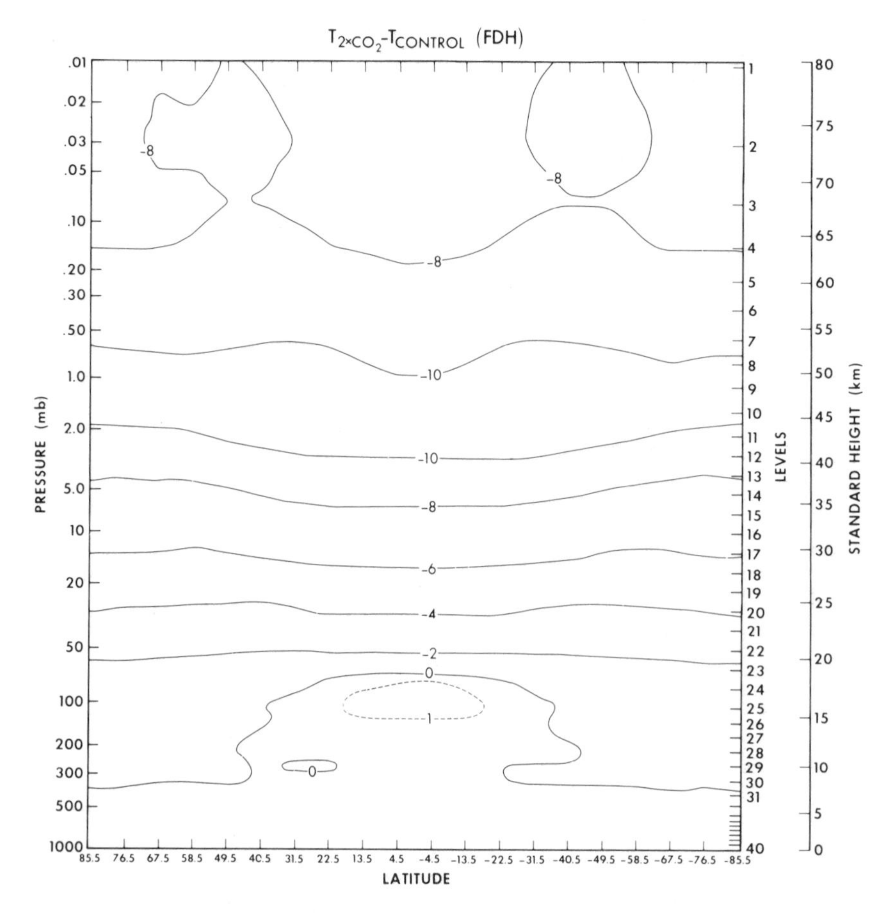

FIGURE 8-2 GFDL "SKYHI" model equilibrium temperature changes due to a doubling of carbon dioxide as determined from an annual mean model with prescribed sea-surface temperatures. On the left is a fixed dynamical heating

upper stratosphere that will be partially compensated for in the lower stratosphere. But, the unpredicted antarctic ozone situation warns us to be wary of things that may be missing from current model calculations. Speculation follows on some of the things that may need to be considered in future models.

The first speculation concerns the possibility of reduced stratospheric transport circulation, that is, reduced efficiency of the meridional circulation. The reason is related to Robert Dickinson's presentation, which follows. The projected greenhouse gas warming in

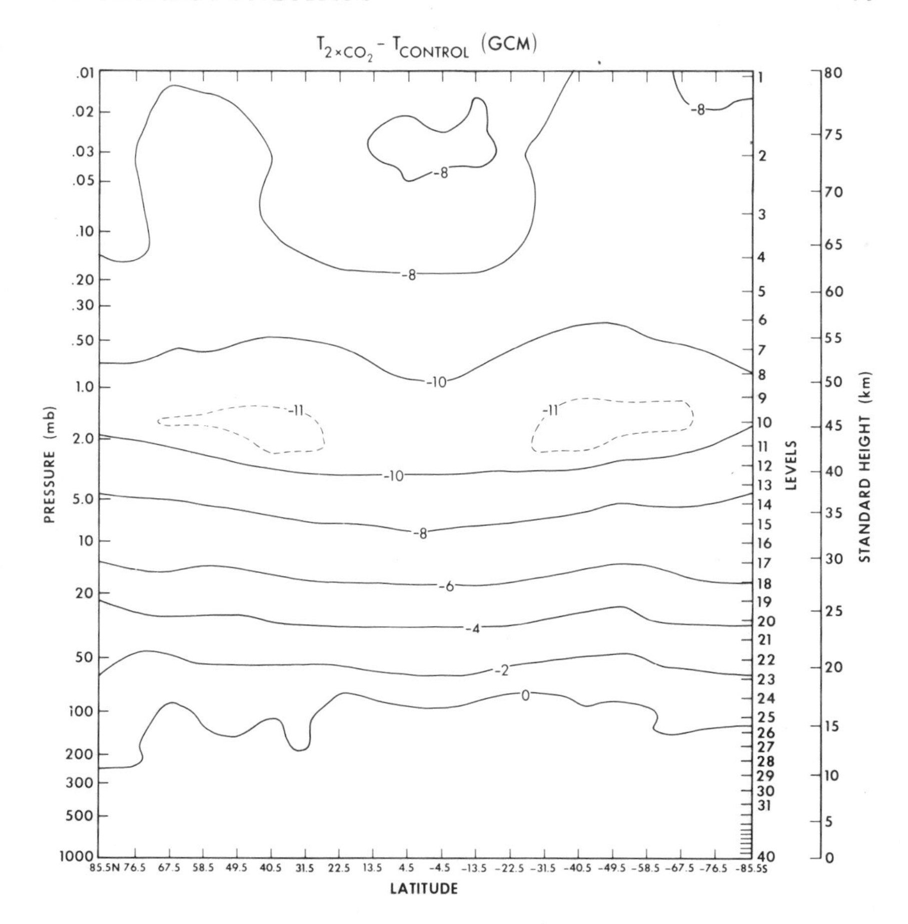

(FDH) calculation; on the right is the full general circulation model (GCM) result. (Reprinted, by permission, from Fels et al., 1980. Copyright © 1980 by The American Meteorological Society.)

the troposphere should result in a weaker meridional temperature gradient, thus weakening tropospheric circulation and decreasing the flux of wave activity to the stratosphere. The stratospheric transport circulation would in turn be reduced.

Another speculation is that water vapor will increase in the lower stratosphere (as well as in the upper stratosphere because of methane). With the column feedback process resulting in increased ozone in the lower stratosphere, it is possible that a significant heating of the equatorial tropopause may occur. A heating of up to 4°C

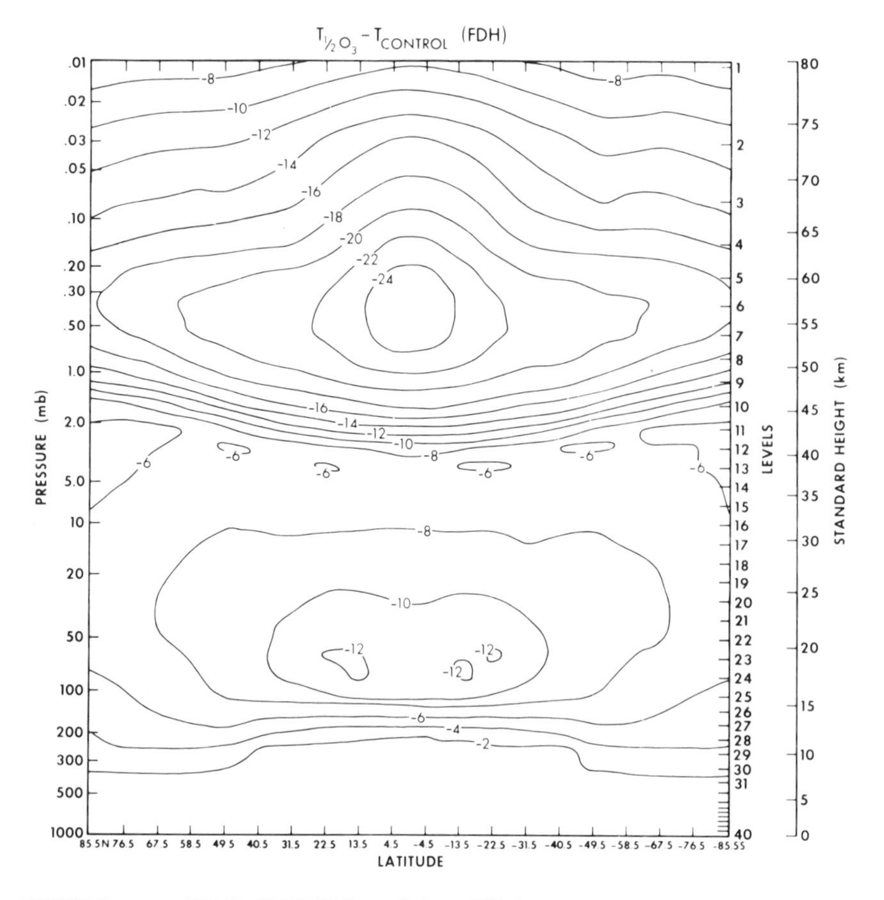

FIGURE 8-3 GFDL "SKYHI" model equilibrium temperature changes due to a uniform 50 percent ozone reduction. On the left is a fixed dynamical heating (FDH) calculation; on the right is the full general circulation model (GCM)

increases the saturation vapor pressure by up to a factor of two. This temperature effect, leading to a water vapor increase in the 15- to 20-km region, may be much greater than that due to methane at these altitudes.

We are beginning to understand the influence of the antarctic ozone seasonal depletion on the ozone climatology of the Southern Hemisphere. It appears that dilution is occurring, causing significant ozone decreases throughout the Southern Hemisphere, as pointed out in the presentations by Robert Watson and F. Sherwood Rowland.

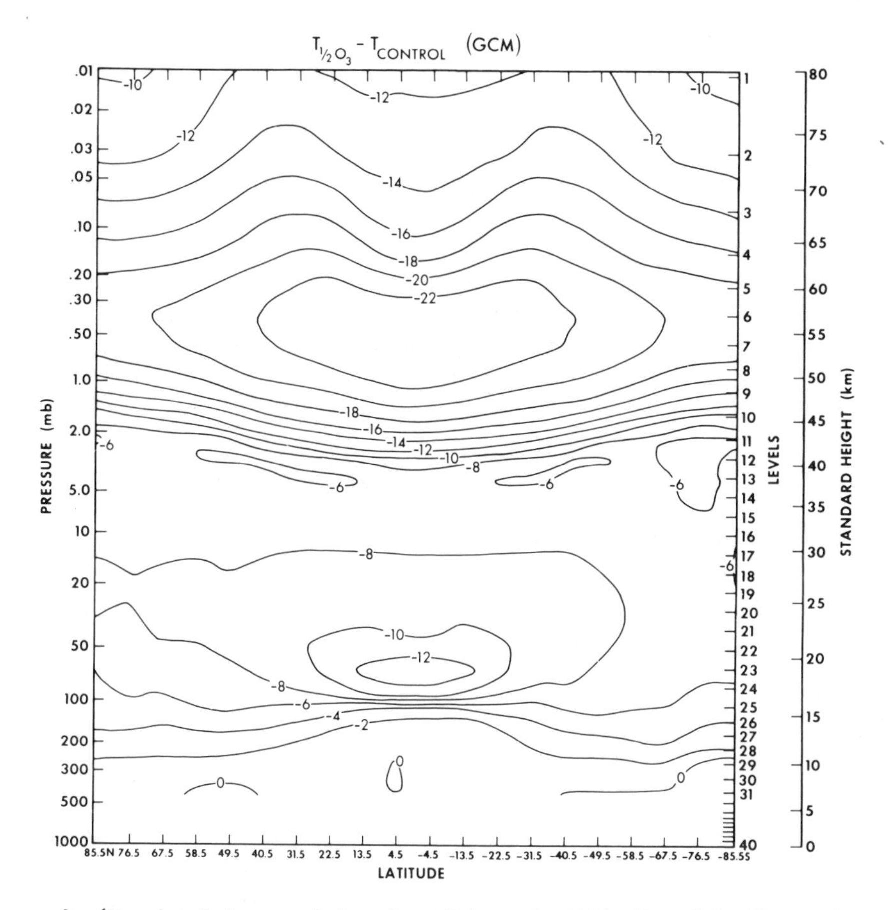

result. (Reprinted, by permission, from Fels et al., 1980. Copyright © 1980 by the American Meteorological Society.)

We currently know much less about the situation in the Northern Hemisphere. However, the northern high-latitude region probably contains more total amount of chlorine species at any particular time in winter than does the southern region in winter, because the Northern Hemisphere region is dynamically more active. Stratospheric cooling due to the combined effects of greenhouse gases and reduced ozone will likely be accented in higher latitudes. We already know of the existence of some polar stratospheric clouds in the northern polar vortex, although they do not form nearly as efficiently there as in the Southern Hemisphere. The reason is not chemical but dynamical; Northern Hemisphere dynamical mixing is greater than that in the Southern Hemisphere. But as the stratosphere cools due to changed greenhouse gases, polar stratospheric clouds should become more common and widespread in the north. Also, reduced transport circulation in the Northern Hemisphere could lead to a further decrease of temperatures at high latitudes. Thus the kind of ozone depletion due to heterogeneous chemistry that is now well documented in the Southern Hemisphere may become common in the Northern Hemisphere as well. I think that much work needs to be done, not only in field observations, but also in theoretical studies and modeling, to quantify this change and the other possible changes that I have mentioned.

(In answer to a question): Only GFDL and NCAR have looked in detail at dynamical modeling of stratospheric climate change. We have learned that the stratosphere has a peculiar nonlinear "switch." This has been noted for many years in sudden warming events, which can temporarily push the stratosphere far out of radiative equilibrium. When this occurs, it seems to be much easier for the system to keep the new configuration, despite the large radiative imbalance. The reason goes back to wave propagation theory, in that if the winds are very strong and the system is very cold, planetary waves are refracted toward the equator with great efficiency. If zonal winds decrease, then the effects of tropospheric forcing are more likely to focus toward higher latitudes. In the climate-modeling case, the models had a cold bias that made model stratospheric receptivity to planetary waves too weak, which in turn accentuated the cold bias. However, if warming is introduced from another cause, then the "switch" may be thrown, rapidly putting the stratosphere into another, quite different, quasi-equilibrium state. Such a switch could also work in the opposite direction. The antarctic polar region, for example, is strongly resistant to any wave forcing because it is on the cold, high-wind-speed side of this implied "dynamical limit." The arctic region, on the other hand, is ***presently*** not so constrained by this dynamical limit because it experiences relatively high levels of dynamical forcing.

(In answer to another question): Rowland, in his talk, and others have speculated that the antarctic ozone hole-generating process may already be operating to some extent in the Northern Hemisphere polar region. However, the effects should be much less noticeable because of the greater dynamic variability there as well as likely smaller levels of ozone chemical destruction.

(In answer to a question about model resolution and model accuracy): Fifteen years ago, we were constrained to work with 9° resolution because of computer and other limitations, but we knew from comparison with observations that we were in big trouble without knowing why. About 8 years ago, we progressed to finer resolution, and results began to reflect much more the real stratosphere. We also found that there were several other modeling problems to contend with. We learned that low-resolution models cannot resolve gravity waves. We also learned that we could sometimes get the right answer even though the radiation was incorrectly specified. (One way to "fix" the cold bias in the low-resolution models was to put in a bad radiation code that allowed us to get closer to the observed temperature.) Only recently have we learned that at least the winter half of the year is dynamical in a fundamental way. There was a tacit assumption that the stratosphere was in radiative equilibrium in the wintertime except during sudden warming events. However, we have learned that tropospheric forcing of the stratosphere does not permit radiative equilibrium to be established, a fact that was first theorized by Dickinson (1975).

To put together a stratospheric model that does not "cheat," in the sense of forcing a lower boundary condition or including a bogus radiation parameterization, requires an extremely long and strong commitment of an interdisciplinary team convinced that spending a decade on the problem is worth it. As a result, there have been few sustained participants in comprehensive stratospheric modeling.

We knew, even 15 years ago, that stratospheric modeling involved more dynamics and three-dimensionality than we were able to represent at the time. My chemical colleagues have challenged me with the following question: What can be said, on the basis of three-dimensional dynamical models, about the viability of one- and two-dimensional models? One thing we did learn is that one- and two-dimensional models do have a theoretically defensible fundamental basis. In both cases, we have learned that the basis is trickier than had been assumed. For example, in one dimension, the eddy diffusion coefficients are a function of the chemistry. Two-dimensional models can capture much of what a three-dimensional model does in a self-consistent way as long as they stick to prescribed transport, but when dynamical adjustment of the stratosphere is a dominant factor, the two-dimensional models are completely inept. Even so, this model hierarchy has great value provided we are also aware of the limitations of each type of model.

Even the 1° dynamical model gives polar temperatures in the lower stratosphere that are a bit too cold. The lower stratosphere is still not dynamical enough, probably because gravity waves are not properly represented. We may have to resort to parameterizing the effects of these gravity waves. However, when a parameterization is introduced to "fix" a model, one is not really justified in perturbing the model climate unless the parameterization is also perturbed accordingly. Here the modeler is faced with a dilemma, since the nature of most parameterizations is that their variation under changed climatic conditions is unknown.

REFERENCES

Dickinson, R.E. 1975. Energetics of the stratosphere. J. Atmos. Terr. Phys. 37:855-864.

Fels, S.B., J.D. Mahlman, M.D. Schwarzkopf, and R.W. Sinclair. 1980. Stratosphere sensitivity to perturbations in ozone and carbon dioxide: Radiative and dynamical response. J. Atmos. Sci. 37:2265-2297.

Mahlman, J.D., and L.J. Umscheid. 1987. Comprehensive modeling of the middle atmosphere: The influence of horizontal resolution. Transport Processes in the Middle Atmosphere, G. Visconti and R. Garcia (eds.), NATO ASI Series C: Mathematical and Physical Sciences, Vol. 213, D. Reidel Publishing Co., 251-266.

Watson, R.T., M.J. Prather, and M.J. Kurylo. 1988. Present State of Knowledge of the Upper Atmosphere 1988: An Assessment Report. NASA Reference Publication No. 1208, National Aeronautics and Space Administration, Washington, D.C.

World Meteorological Organization-National Aeronautics and Space Administration (WMO-NASA). 1986. Atmospheric Ozone 1985: Assessment of Our Understanding of the Processes Controlling Its Present Distribution and Change. Global Ozone Research and Monitoring Project, Report No. 16, 3 vols., WMO, Geneva.

9
Historical Trends in Atmospheric Methane Concentration and the Temperature Sensitivity of Methane Outgassing from Boreal and Polar Regions

ROBERT C. HARRISS
Langley Research Center
National Aeronautics and Space Administration

Recent studies have documented two trends in atmospheric methane (CH_4) concentrations. First, a modern trend of increasing global atmospheric CH_4 has been documented in the trapped gas in polar ice cores (Craig and Chou, 1982; Khalil and Rasmussen, 1987) and with regular monitoring of ambient CH_4 at remote locations around the world (Steele et al., 1987; Blake and Rowland, 1988). These data indicate that CH_4 has increased from a concentration of approximately 650 parts per billion by volume (ppbv) 200 years ago to 1,690 ppbv in 1988 (Figure 9-1). This recent increase in atmospheric CH_4 over the past several hundred years correlates with the growth of the human population and industrial society and is hypothesized to be a result of increased CH_4 emissions related primarily to the expansion of rice agriculture, domestication of ruminant animals, landfilling of organic wastes, and the mining and use of fossil fuels (Ehhalt, 1985; Pearman and Fraser, 1988).

A second trend from low concentrations of CH_4 (350 ppbv) at times of glacial maximum (approximately 20,000 years B.P.), increasing to 650 ppbv during interglacial times (Figure 9-2) has been observed in ice core samples from Antarctica and Greenland (Stauffer et al., 1988). This variability in prehistoric CH_4 is hypothesized to be due to the expansion of arctic and boreal peatlands following glacial retreat (Harriss et al., 1985). These high-latitude ecosystems are commonly wetlands with oxygen-deficient, organic-rich soils that

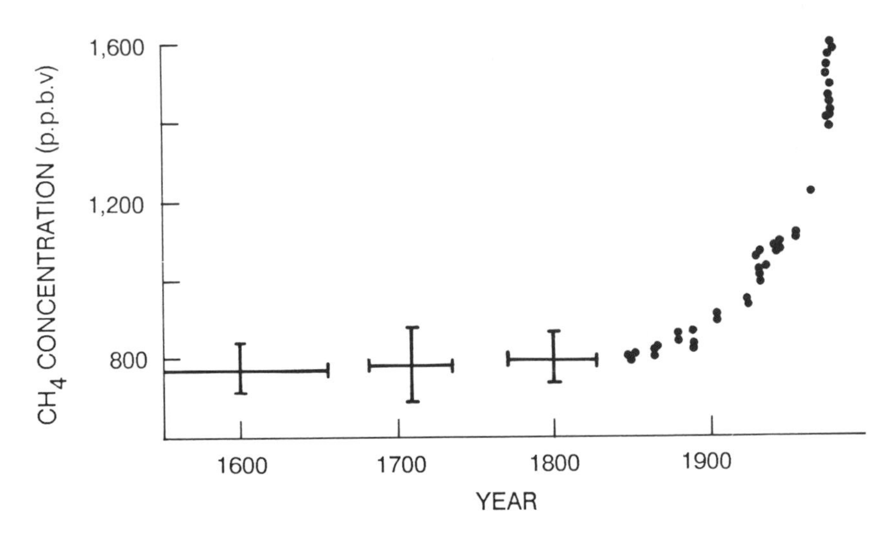

FIGURE 9-1 Atmospheric methane (CH_4) variations (in ppbv) over the past few centuries. Each point represents measurements or (in the earlier centuries) estimates with error bars. (Adapted from Khalil and Rasmussen, 1987.)

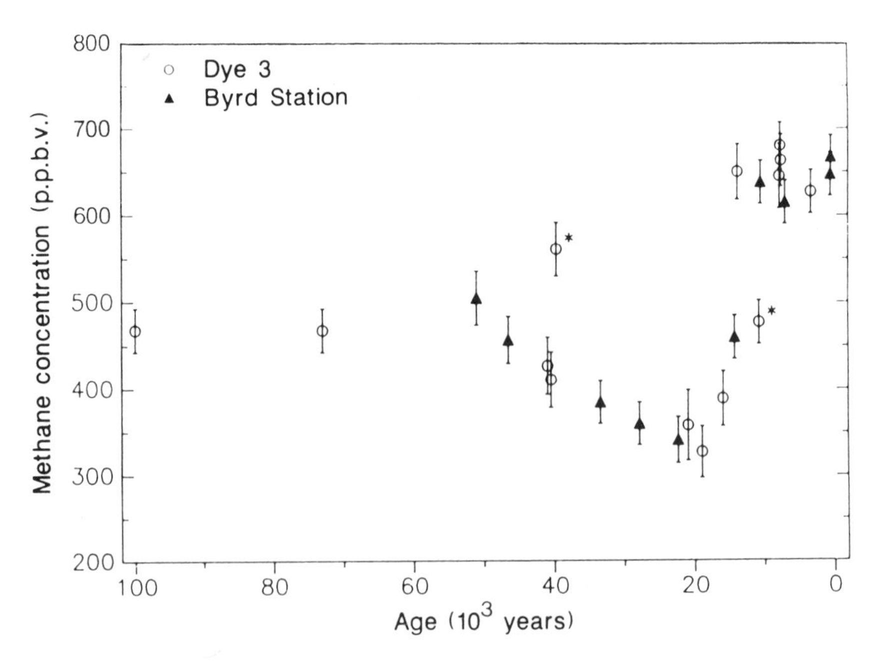

FIGURE 9-2 Atmospheric methane (CH_4) variations in glacial and interglacial times, as determined from ice corings. Abscissa is thousands of years before present. (Reprinted, by permission, from Stauffer et al., 1988. Copyright ©1988 by Macmillan Magazines Ltd.)

emit CH_4 as a byproduct of microbial decomposition processes. It has been estimated that these boreal and arctic peatlands may currently produce about 60 percent of the global CH_4 flux to the atmosphere from natural wetlands (Matthews and Fung, 1987).

Methane is also present in large quantities as "frozen" CH_4 clathrates in subsurface sediments of the polar regions, and as trapped gas in permafrost (Bell, 1982; Revelle, 1983). If substantial warming of the polar oceans and landscape were to occur, these "frozen" sources of CH_4 would be released to migrate through the soil and overlying sediments into the atmosphere. However, at present this issue is highly speculative.

There is little doubt that increasing concentrations of CH_4 and other trace gases can have a profound influence on the earth's atmospheric chemistry and on climate (Thompson and Cicerone, 1986; Ramanathan, 1988). The question of what influence climate change has on emissions of trace gases from the global biosphere is less certain. If emissions of CH_4 are enhanced by global warming, a positive feedback can result, with the increasing concentrations of atmospheric CH_4 further enhancing the tendency for a greenhouse warming. Such positive feedback could contribute to abrupt climate changes, resulting in considerable ecological and societal disruption. At present much less is known about negative feedbacks on climate warming such as increased evaporation, which would lead to more clouds and a higher albedo, with subsequent cooling of the earth's surface. Increased atmospheric water vapor could also lead to higher concentrations of atmospheric hydroxyl (OH), which destroys CH_4, consequently mitigating the increase in source emissions due to climate warming.

Very little quantitative information is currently available on the long-term, integrated response of sources or sinks of atmospheric CH_4 to climate change. The data that are available on the temperature sensitivity of CH_4 sources from organic soils and sediments clearly indicate that CH_4 emission rates increase with increasing surface temperature (Baker-Blocker et al., 1977; Crill et al., 1988). Figure 9-3 illustrates the response of CH_4 emissions to seasonal warming of boreal peatland soils in northern Minnesota. From these data it is reasonable to hypothesize that the *initial* response to warming of both natural and anthropogenic organic soils and sediments, which are the dominant sources of atmospheric CH_4, will be an increasing flux of CH_4 to the atmosphere. A large fraction of the world's old carbon is in boreal and arctic regions, so these ecosystems are of

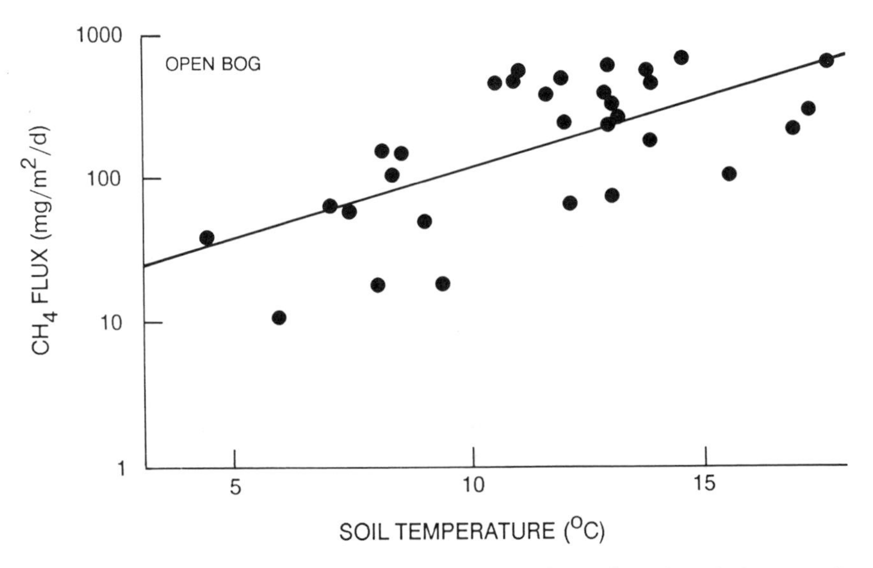

FIGURE 9-3 Effect of soil temperature on methane (CH_4) emissions to the atmosphere from an open bog in the Marcell Experimental Forest, Minnesota.

particular importance to understanding the response of CH_4 sources to climate change. Thus, if the greenhouse warming predicted by models (e.g., see Ramanathan, 1988) is realized, it is likely that the rate of increase in atmospheric CH_4 will increase further over the next few decades due to enhanced flux from boreal and arctic wetlands, rice paddy soils, landfills, and other soil and sediment sources.

The mid-term (10- to 100-year) response of CH_4 sources to a greenhouse warming of the earth's surface is impossible to predict at present. Major scientific issues related to negative feedback mechanisms on the CH_4 increase (e.g., increases in atmospheric OH and drying of wetland soils in major source regions) must be resolved.

Recent advances in environmental measurement technologies and techniques make possible, for the first time, regional- to global-scale quantification of biosphere-atmosphere interactions. The uncertainties in how CH_4 sources and sinks will respond to future climate change or to socioeconomic developments that influence CH_4 sources can be reduced by vigorous research programs in global tropospheric chemistry (NRC, 1984) and earth system science (NRC, 1985). Specific research to better understand the importance of increasing atmospheric CH_4 as a cause and/or consequence of climate change must include the following components:

1. The quantification of both natural and anthropogenic CH_4 sources must be improved. Such improvements will require integrated ground, aircraft, and satellite measurements, which will provide accurate estimates of CH_4 flux to the atmosphere at regional scales. Isotope studies using newly developed accelerator mass spectrometric techniques (Lowe et al., 1988), combined with more detailed temporal resolution from ice core sampling, could resolve the issue of what contribution fossil sources of CH_4 make to the variability in atmospheric CH_4 over long and short time scales.

2. The boreal and arctic regions store much of the earth's soil carbon in wetlands, which are important sources of atmospheric CH_4. It will be especially important to quantify the temperature sensitivity of the physical and biological processes responsible for CH_4 outgassing from these ecosystems, since they are in regions where climate models predict an enhanced greenhouse warming effect.

3. The impact of increasing concentrations of tropospheric carbon monoxide, ozone, and CH_4 on global OH distributions can be resolved with the availability of quantitative source data and the development of advanced three-dimensional photochemical models for prediction of OH.

4. The flux of CH_4 from the troposphere to the stratosphere can be quantified with a comprehensive program of aircraft and satellite measurements. Methane decomposes in the stratosphere to products such as water vapor that can alter chemical and radiative transfer processes.

In summary, during the past decade it has become clear that current changes in the earth's atmospheric composition are global in scale. Studies on atmospheric CH_4 will provide critical information on how the earth's biosphere and atmosphere will respond to the global warming that is forecast by climate models to occur in the next few decades. Such research will also provide the necessary scientific data to make sound regulatory decisions if policymakers decide to arrest or reverse the growth of sources of CH_4 produced by human activities.

REFERENCES

Baker-Blocker, A., T.M. Donahue, and K.H. Mancy. 1977. Methane flux from wetland areas. Tellus 29:245-250.

Bell, P.R. 1982. Methane hydrate and the carbon dioxide question. In Carbon Dioxide Review 1982, W.C. Clark (ed.). Oxford University Press, New York, pp. 401-406.

Blake, D.R., and F.S. Rowland. 1988. Continuing worldwide increase in tropospheric methane, 1978 to 1987. Science 239:1129-1131.

Craig, H., and C.C. Chou. 1982. Methane: Record in polar ice cores. Geophys. Res. Lett. 9:1221-1224.

Crill, P.M., K.B. Bartlett, R.C. Harriss, E. Gorham, E.S. Verry, D.I. Sebacher, L. Madzar, and W. Sanner. 1988. Methane flux from Minnesota peatlands. Global Biogeochem. Cycles (in press).

Ehhalt, D.H. 1985. Methane in the global atmosphere. Environment 27:6-12.

Harriss, R.C., E. Gorham, D.I. Sebacher, K.B. Bartlett, and P.A. Flebbe. 1985. Methane flux from northern peatlands. Nature 315:652-653.

Khalil, M.A.K., and R.A. Rasmussen. 1987. Atmospheric methane: Trends over the last 10,000 years. Atmos. Environ. 21:2445-2452.

Lowe, D.C., C.A.M. Brenninkmeijer, M.R. Manning, R. Sparks, and G. Wallace. 1988. Radiocarbon determination of atmospheric methane at Baring Head, New Zealand. Nature 332:522-524.

Matthews, E., and I. Fung. 1987. Methane emission from natural wetlands: Global distribution, area, and environmental characteristics of sources. Global Biogeochem. Cycles 1:61-86.

National Research Council (NRC). 1984. Global Tropospheric Chemistry: A Plan for Action, National Academy Press, Washington, D.C., 194 pp.

National Research Council (NRC). 1985. A Strategy for Earth Science from Space in the 1980's and 1990's, Part II: Atmosphere and Interactions with the Solid Earth, Oceans, and Biota, National Academy Press, Washington, D.C., 149 pp.

Pearman, G.I., and P.J. Fraser. 1988. Sources of increased methane. Nature 332:489-490.

Ramanathan, V. 1988. The greenhouse theory of climate change: A test by an inadvertent global experiment. Science 240:293-299.

Revelle, R.R. 1983. Methane hydrates in continental slope sediments and increasing atmospheric carbon dioxide. Pp. 252-261 in Changing Climate, National Academy Press, Washington, D.C.

Stauffer, B., E. Lochbronner, H. Oeschger, and J. Schwander. 1988. Methane concentration in the glacial atmosphere was only half that of the preindustrial Holocene. Nature 332:812-814.

Steele, L.P., P.J. Fraser, R.A. Rasmussen, M.A.K. Khalil, T.J. Conway, A.J. Crawford, R.H. Gammon, K.A. Masarie, and K.W. Thoning. 1987. The global distribution of methane in the troposphere. J. Atmos. Chem. 5:125-171.

Thompson, A.M., and R.J. Cicerone. 1986. Possible perturbations to atmospheric CO, CH_4, and OH. J. Geophys. Res. 91:10853-10864.

10
Global Temperature Trends

KEVIN E. TRENBERTH
National Center for Atmospheric Research

This talk focuses mainly on stratospheric temperature trends but covers some aspects of tropospheric temperatures as well. The last part of the talk discusses what is happening in the antarctic region. The principal issue to consider is the consistency of ozone and other trace gas changes, and of the expected temperature changes resulting from such trace gas changes, with actual temperature changes. In the lower stratosphere, especially during the polar night, temperature and ozone are similarly advected, so from that standpoint temperature and ozone changes should occur in tandem. But ozone changes also imply changes in solar heating rates and thus changes in temperature. Are changes in ozone and temperature consistent in that sense?

With regard to greenhouse gases, theory indicates that a warming should occur at the surface, but since the oceans and the cryosphere are involved, changes are likely to occur on time scales of a decade or more, whereas the response of the stratosphere to greenhouse gases should occur on a rather short time scale. For the stratosphere, ozone decrease and greenhouse gas increase both imply cooling; therefore the chance of detecting an undoubted change is greater than in the troposphere. Jerry Mahlman has already pointed out that comparatively large, unambiguous changes have occurred in the upper stratosphere, but elsewhere temperature trends are much harder to find.

One of the problems with temperature trends is the difficulty of detecting them. Temperature records are clearly functions of the measurement systems used. The issue of temperature measurements needs to be addressed in the future, especially with regard to any climate change program. The main device for taking temperature measurements in the atmosphere is the radiosonde, which was not, however, designed to detect long-term temperature trends. There have been changes in type and design (especially with regard to radiation shields), and significant day-night temperature differences that have varied with design have resulted. A recent international intercalibration found that instruments of different nations generally reproduce temperatures within about 0.2°C and pressures within 2 mb. This amount of error is not serious in the troposphere but becomes significant in the stratosphere, where 2 mb is a sizeable fraction of the pressure being measured. Hence, radiosondes are not useful above about 20 mb; in fact, few of them ever rise higher than 20 mb.

The other principal source of temperature data in the free atmosphere is satellite radiance data that exist in useful form (as TOVS data) from about 1979 to present. There are again problems with calibration, changes to new satellites, and changes in the time of day of satellite overhead passage. These problems have led to temperature differences of 2 to 4°C, due to changes of satellite in the stratosphere especially. Therefore, it has been necessary to adjust and merge segments of the satellite record in order to get a continuous record of temperature trends. Some of the analyses have in fact removed apparent (but spurious) trends by invoking this kind of step-function adjustment. So it is very difficult to get an absolute calibration and to use satellite radiance data to get reliable trends.

Rocketsondes are too few and far between to be useful for establishing trends alone, but they may be useful for calibrating satellite data in the region above the useful radiosonde ascents. But rocketsondes have also changed, in type and instrumentation, with time.

The products that exist for determining temperature trends are station or instrument records and analyses based in some way on these records. Many station records suffer from missing data, a major problem. Additionally, the areal coverage is insufficient in many regions. Various kinds of analyses have been done to fill the record gaps and to merge different data types (e.g., Labitzke et al. (1986) have compiled temperature analyses for the Northern Hemisphere that start in 1965).

When one examines a temperature record for a linear trend, one needs to be concerned with "end effects" and with major temporary perturbations, such as those attributed to El Chichon. The seasonal cycle, which has very large amplitude in the stratosphere, is important in that incomplete data during a period of rapid seasonal change can give misleading results, as explained below. The missing data problem arises because of balloon loss or data transmission problems. Because of differing data amounts at different levels, the thickness of the monthly mean geopotential heights may not equal the monthly mean of the daily thickness values.

The effects on temperature trends that we need to consider include the solar cycle with its variation in ultraviolet flux, leading to stratospheric cooling at the solar minimum; aerosols from volcanic eruptions, resulting in heating within the stratospheric aerosol layer; ozone change, with a decrease implying stratospheric cooling; carbon dioxide, with an increase implying stratospheric cooling; and natural variability due to dynamical processes and internally generated variations such as the quasi-biennial oscillation, southern oscillation, and sudden stratospheric warmings, affecting both temperatures and the ozone distribution.

The surface temperature record (Figure 10-1) as compiled by Hansen and Lebedeff (1988) shows that, in terms of global averages, 1987 was the second warmest year on record, being exceeded only by 1981. The 1980s thus far average distinctly warmer than any previous decade since the record began in the 1880s. It is noteworthy that global temperatures have been highest during a period of decrease in insolation as measured by satellite. The magnitude of the insolation decrease since 1979 is estimated to have offset almost exactly the expected increase since 1979 in temperature due to greenhouse gases. The trend of the 108-year record is not linear but has variations, including a relatively warm period in the 1930s, which has been attributed to a dearth of volcanic aerosols but could also be due to natural variability. There is a problem in interpreting this record because of such anthropogenic effects as development of urban heat islands (Kukla et al., 1986), deforestation, building of roads, and relocation of observing stations. Although an attempt has been made to estimate such effects, we cannot be sure that the record is indicating changes due to carbon dioxide buildup and other external factors.

As I mentioned earlier, problems can arise in fitting linear trends to data. One can use least squares regression to fit a linear trend to

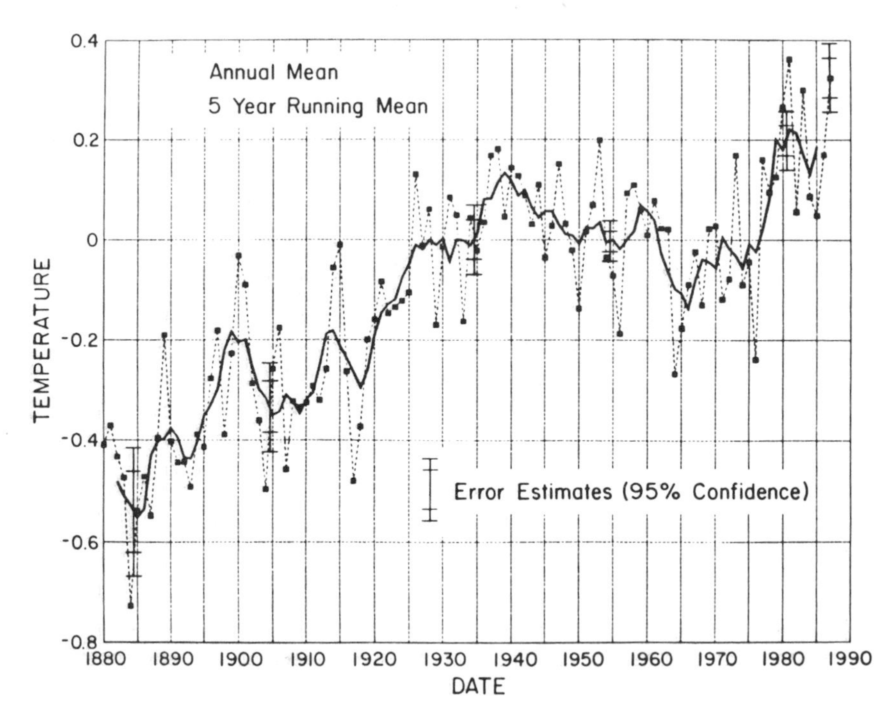

FIGURE 10-1 Departure of mean global temperatures from their 1951 to 1980 period mean value for individual years (dots connected with dashed lines) and for 5-year running mean values (solid curve). Error estimates for both individual and running mean values are shown for selected years. Period of record is 1880 through 1987. (Reprinted from Hansen and Lebedeff, 1988. Copyright © 1988 by the American Geophysical Union.)

one cycle (from pi to three pi) of a sine wave and get a large positive trend with a correlation of 0.78, which is highly significant. If half a cycle is added to the beginning of this record and another half to the end, a large downward linear trend results that has a correlation of −0.39. The point of this is that trends are very sensitive to the beginning and end points of the record. Conclusions about what trends exist are greatly influenced by end effects when the period of record is comparatively short. The year 1979 turns out to be a bad year to start a record but has been used in many stratospheric satellite records. Several upper-level tropospheric records start around 1964, which, as seen from Figure 10-1, has the lowest surface temperature value since the 1920s. Any trend starting with 1964 is likely to be totally unrepresentative of longer-period trends.

Karoly (1987) has looked at temperature trends for 19 Southern Hemisphere stations from 1964 to 1985. He found slight upward trends, ranging from 0.1 to 1.2°C, at 700 mb at all of these stations and slight downward trends, ranging from −0.1 to −1.3, at 50 mb at all but one of them. The Mt. Agung volcanic eruption that occurred in 1963 may have affected the trends to some extent; hence the trends may reflect more than just the carbon dioxide increase. Also in the Southern Hemisphere, Salinger (1979) has composited New Zealand mean annual and seasonal surface temperatures from 1853 through 1975. A minimum in the mid-1960s is even more pronounced in this record than in the global record of Hansen and Lebedeff (1988), and a marked upward trend since about 1964 is not representative of the overall long-term trend of this record.

Labitzke et al. (1986) have published time-series of monthly mean 30-mb temperatures for the Northern Hemisphere from 1965 through 1983. The El Chichon eruption of 1982, and perhaps the Mt. St. Helens eruption of 1980, contaminated the record in the last few years. Therefore Labitzke computed linear trends from 1966 through 1980. These trends are slightly downward for this period at latitudes poleward of 20°N, but the time series show much interannual variability, even though the quasi-biennial oscillation has been removed from these series. Only the trend at 70°N latitude appears large enough to be meaningful.

Using a relatively sparse network of stations, Angell (1986a) has looked at global temperature trends from 1960 to 1985. His results show a slight warming in the Southern Hemisphere midtroposphere and a slight cooling above about 10 km, in general agreement with Karoly's results. These trends are somewhat stronger for the decade 1975 to 1985 alone. More recent results of Angell (unpublished) show layer mean temperature time-series for the world through 1987. The records for the surface, the 850- to 300-mb layer, and the 300- to 100-mb layer start in 1957, and the 100- to 50-mb record starts in 1969. The surface and tropospheric layer both show cooling until about 1964 to 1965 and a warming trend since. The 300- to 100-mb layer does not show a recognizable trend. At 100 to 50 mb, Angell's data show a pronounced downward trend after a warm peak in 1982 that was probably associated with the eruption of El Chichon. Angell (1986b) did a latitudinal breakdown of the 100- to 50-mb data, which shows that a recent rapid decline of 7°C in the South Polar region accounts for much of the globally averaged decline, with the equatorial zone also making a significant contribution.

The Ozone Trends Panel (Watson et al., 1988; Trenberth, 1988) did a comparison of temperature time-series by Angell (1986b; 100 to 50 mb), Labitzke et al. (1986; 100 to 50 mb, for the Northern Hemisphere only), the National Meteorological Center (NMC) (70 to 50 mb, based on NMC analyses), and the microwave sounder unit (approximately 90 mb). There is good agreement for the Northern Hemisphere from 1979 through 1986, showing a warm peak in 1982 to 1983 that is probably due to El Niño or El Chichon, followed by a slight cooling trend. In the Southern Hemisphere, Angell's values indicate much colder temperatures than those for the other two series in 1985 to 1986. Angell's series also indicates the coldest temperatures in recent years in the 30°N to 30°S latitude region. A similar comparison for the 5- to 1-mb layer, based on the NMC analysis and two SSU channels, shows a clear downward trend in temperatures from 1979 to 1986 for all major regions of the world.

For the antarctic region, time-series from 1979 to 1986 of 200-mb heights generated in analyses by the NMC and the European Centre for Medium-Range Weather Forecasting (ECMWF) analyses disagree by 50 to 400 m (Trenberth and Olson, 1988). The observational input to these analyses has an accuracy of about 20 m, for comparison. The ECMWF analysis has been getting progressively better with time. The NMC values were highly erroneous between 1983 and 1986 but improved in May 1986, when a correction to the analysis procedure was made. Clearly, the operational analyses in the antarctic region need further improvement.

The record of radiosonde ascents at the Amundsen-Scott South Polar station has been examined and reveals that for the period 1961 through 1976, about 80 to 85 percent of ascents reached the 100-mb level during the winter (June to August) months, with about 40 percent surviving to 50 mb. For the period 1976 to 1986, on the other hand, only about 15 percent of these soundings reached 100 mb or higher, and only about 10 percent managed to reach 50 mb. The reason for this loss is the bursting of balloons in cold temperatures, so that a warm bias is likely to result from the remaining observations that are actually made. The situation is even a little worse at the McMurdo station. The worst month, in terms of the number of observations made, is August, but by October the amount of high-altitude radiosonde data obtained is much greater. This problem has a marked impact on the temperature records.

If we look at the 500-mb, 100-mb, 50-mb, and 30-mb annual cycles of temperature at Amundsen-Scott (Figure 10-2), we note that there is a very steep increase in temperature from August to October. At 50 mb, the average temperature is −80°C at the beginning of October and −55°C at the end of the month. If there is a preponderance of missing data at the beginning of the month, then the monthly mean temperature will have a warm bias. To date, analysis of the records of antarctic stations has not taken this problem into account. As a result, trends in temperatures deduced from antarctic station data are of questionable value.

We have computed the departure from the mean annual cycle (Figure 10-2) of every individual day and then compiled these anomalies into monthly means for the antarctic stations. For 50 mb at McMurdo Sound, this compilation, when compared with straight averages of the available observations for each month, gives monthly means that differ by up to 12°C for some Octobers. For recent years, our method gives mean October temperatures that are consistently up to 8°C colder than those obtained by the straight averaging method. The adjusted October record since 1956 shows a slight downward trend that is not apparent in the unadjusted record.

The above procedure was used for 30-, 50-, and 100-mb data at both Amundsen-Scott and McMurdo to correct for the annual cycle-missing data effect (see Figures 10-3 and 10-4). For the month of September, there is little apparent trend from 1961 to 1987 at Amundsen-Scott or from 1956 to 1986 (with some years missing) at McMurdo. For October (Figure 10-3), a downward trend, more pronounced at Amundsen-Scott, is apparent in the last 8 to 10 years of these records, with the coldest years occurring near the end of the record. Even so, the recent downturn is modest and would probably not be considered especially noteworthy were it not for the known springtime ozone depletion in the antarctic region. For November (Figure 10-4), the recent downward trend is again apparent at Amundsen-Scott for 50 and 100 mb and at McMurdo for 100 mb only. It is noteworthy, however, that at McMurdo the coldest November temperatures at all levels occurred in 1959 and 1961 (1960 data are missing). For 1987, we have recently estimated the anomalies, shown by dashes in Figures 10-3 and 10-4, at the South Pole, and the lowest values on record occur in October and especially in November, consistent with the absence of heating due to the ozone hole.

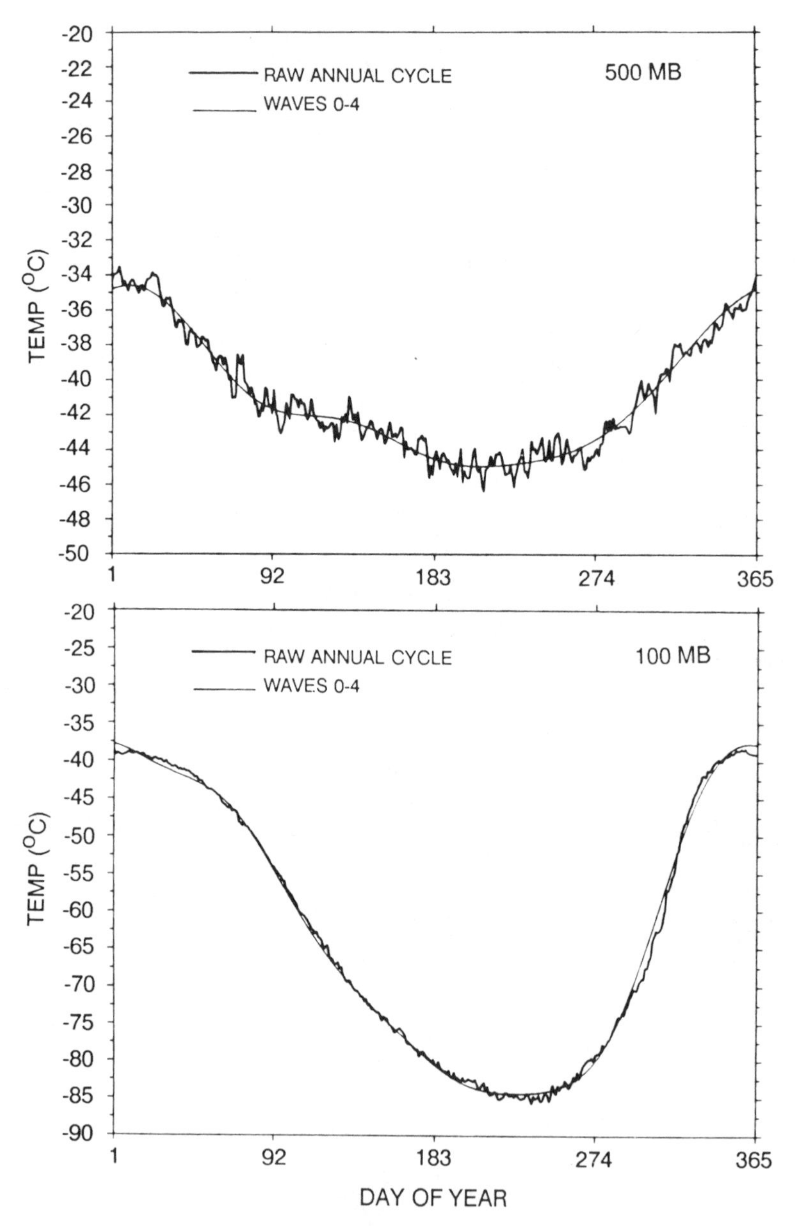

FIGURE 10-2 Mean annual cycle of temperatures at four pressure levels as observed at the Amundsen-Scott antarctic station. Heavy curve shows the daily raw means; thin curve is the best fit of combined temporal waves 0 to 4 to the daily means. Period of record is 1961 through 1986.

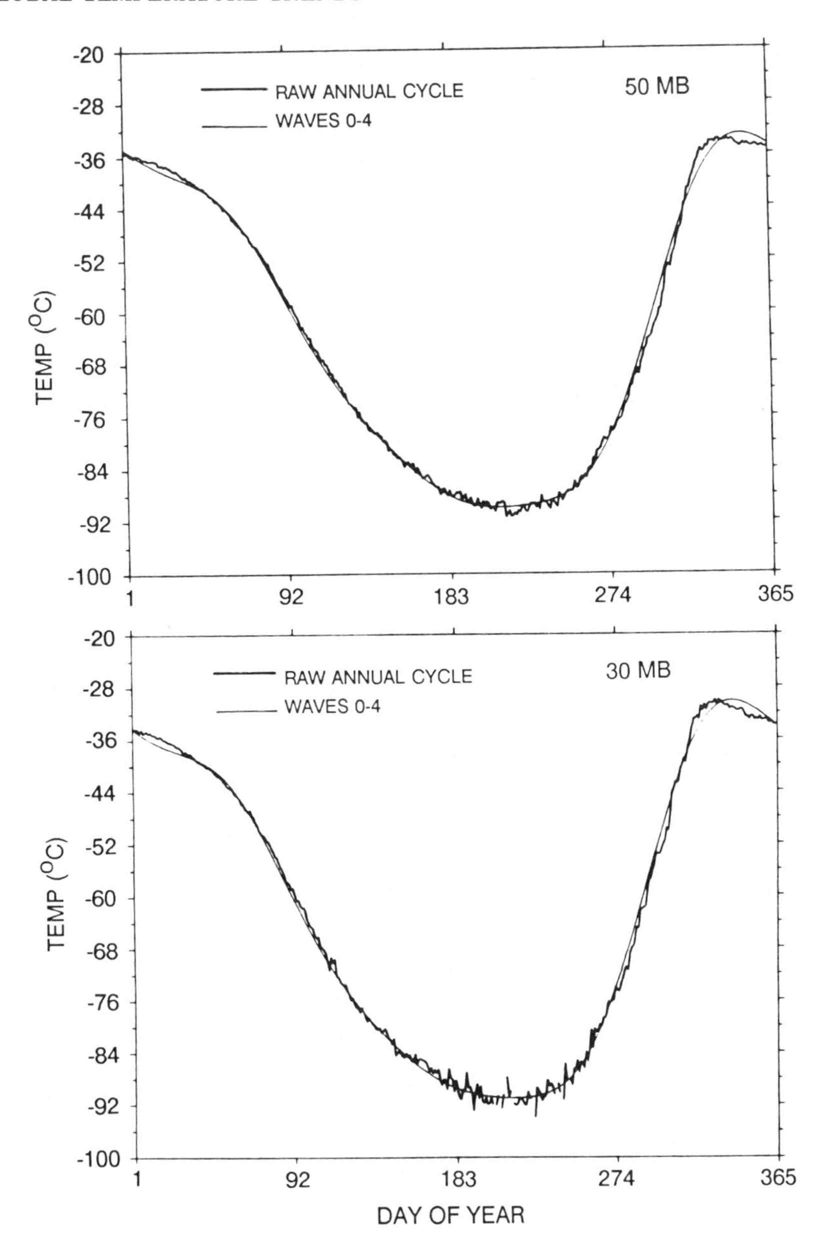

FIGURE 10-2 (continued).

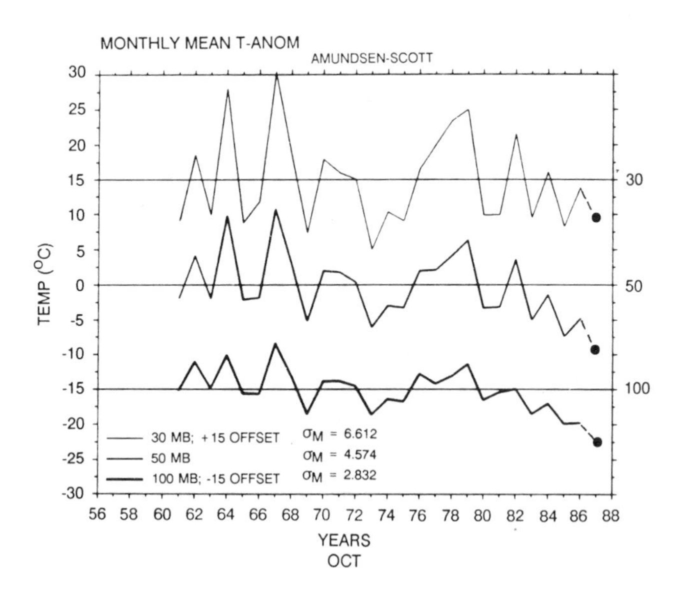

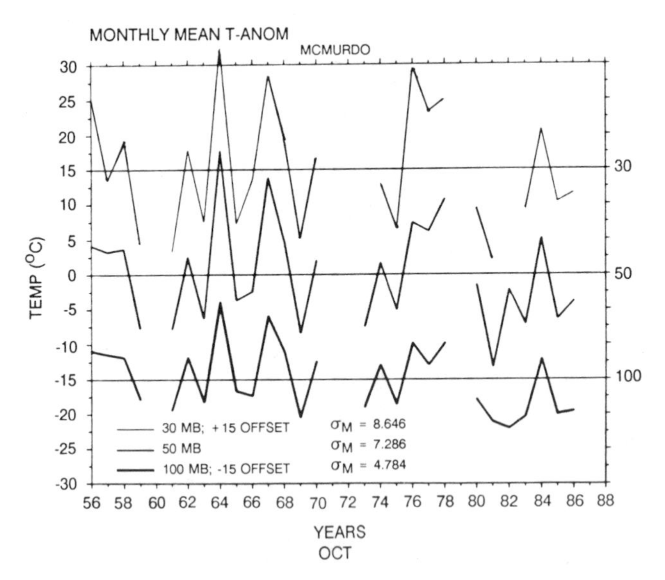

FIGURE 10-3 Monthly mean departure of daily temperature from the period normal for three pressure levels at Amundsen-Scott (top) and McMurdo (bottom) for October. Note that curves for the three different levels are offset from one another by 15°C intervals. Data are missing for some years at McMurdo. The 1987 values for Amundsen-Scott are preliminary. See text for further explanation.

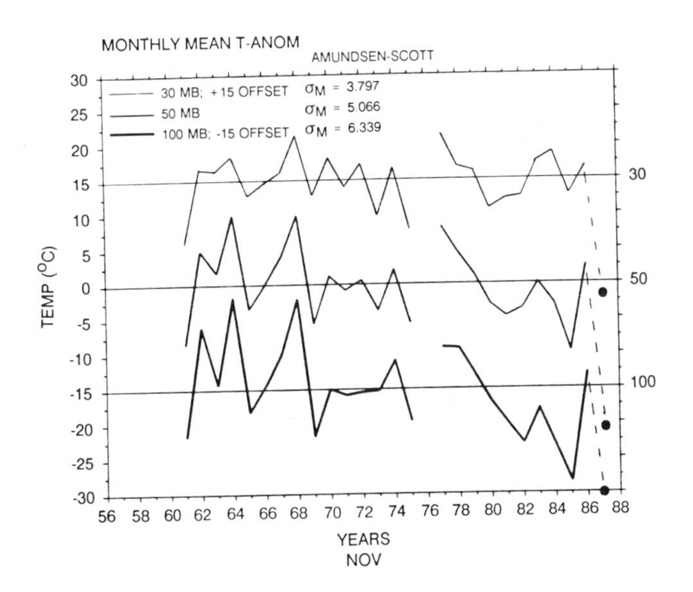

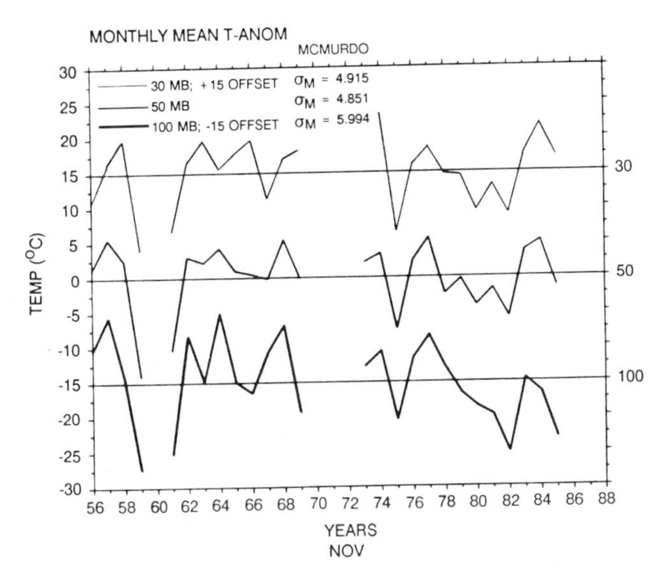

FIGURE 10-4 Monthly mean departure of daily temperature from the period normal for three pressure levels at Amundsen-Scott (top) and McMurdo (bottom) for November. Note that curves for the three different levels are offset from one another by 15°C intervals. Data are missing for some years at McMurdo. The 1987 values for Amundsen-Scott are preliminary. See text for further explanation.

Finally, the Japanese antarctic station Syowa has measured both 100-mb temperatures and total column ozone from 1966 through 1985 (Chubachi, 1986). For October and November, a distinct downward trend in both temperature and ozone is apparent after about 1978. December data show a slight downward trend in recent years. The year-to-year variations in temperature and ozone are approximately parallel to each other, supporting the argument that ozone is the key factor affecting temperatures at this level through solar heating. Syowa 50-mb November temperatures also show a downward trend from about 1978 on. Syowa is located close to the region where the largest trends are supposed to be occurring.

(In answer to a question): The global surface temperature trends that have been computed are weighted toward the Northern Hemisphere, especially for the earlier years. Surface data coverage in the Southern Hemisphere before World War II was only about half the present coverage, and there were no useful antarctic data prior to about 1954.

REFERENCES

Angell, J.K. 1986a. Annual and seasonal global temperature changes in the troposphere and low stratosphere, 1960-1985. Mon. Wea. Rev. 114:1922-1930.

Angell, J.K. 1986b. The close relation between Antarctic total-ozone depletion and cooling of the Antarctic low stratosphere. Geophys. Res. Lett. 13:1240-1243.

Chubachi, S. 1986. On the cooling of stratospheric temperature at Syowa, Antarctica. Geophys. Res. Lett. 13:1221-1223.

Hansen, J., and S. Lebedeff. 1987. Global trends of measured surface air temperature. J. Geophys. Res. 92:13345-13372.

Hansen, J., and S. Lebedeff. 1988. Global surface temperatures: Update through 1987. Geophys. Res. Lett. 15:323-326.

Karoly, D.J. 1987. Southern Hemisphere temperature trends: A possible greenhouse gas effect? Geophys. Res. Lett. 14:1139-1141.

Kukla, G., J. Gavin, and T.R. Karl. 1986. Urban warming. J. Climate Appl. Meteorol. 25:1265-1270.

Labitzke, K., G. Brasseur, B. Naujokat, and A. De Rudder. 1986. Long-term temperature trends in the stratosphere: Possible influence of anthropogenic gases. Geophys. Res. Lett. 13:52-55.

Salinger, M.J. 1979. New Zealand climate: The temperature record, historical data and some agricultural implications. Climatic Change 2:109-126.

Trenberth, K.E. 1988. Review of "Executive summary of the Ozone Trends Panel Report." Environment 30:25-26.

Trenberth, K.E., and J.G. Olson. 1988. An evaluation and intercomparison of global analyses from the National Meteorological Center and the European Centre for Medium Range Weather Forecasts. Bull. Am. Meteorol. Soc. 69:1047-1057.

Watson, R.T., M.J. Prather, and M.J. Kurylo. 1988. Present State of Knowledge of the Upper Atmosphere 1988: An Assessment Report. NASA Reference Publication No. 1208. National Aeronautics and Space Administration, Washington, D.C.

11
Use of Numerical Models to Project Greenhouse Gas-Induced Warming in Polar Regions (The Conceptual Basis Developed Over the Last Twenty Years)

ROBERT E. DICKINSON
National Center for Atmospheric Research

The development of numerical models that will accurately predict climatic change caused by greenhouse gases has been a long-term, continuing research effort (Bolin et al., 1986). The focus of this talk will be the concepts involved and the results of that effort, especially as they apply to the polar regions. There are several major points to consider; I will list them briefly here and then return to each of them later in the talk. The major points are as follows:

1. Expected tropospheric temperature increases from greenhouse gases will be accompanied by temperature decreases, on a different time scale, in the stratosphere.
2. Classical (postulated before 1974) mechanisms of high-latitude amplification by means of positive feedback processes involving the temperature lapse rate and planetary albedo are important.
3. The projected magnitude of high-latitude warming is strongly dependent on the seasonal cycle.
4. Horizontal transport of heat by both the atmosphere and oceans has an important effect on high-latitude temperatures.
5. Vertical energy transfer, as modeled by convective parameterization, also has a significant role for high-latitude temperatures (although not as large a role as in low latitudes).
6. High-latitude cloud cover has a key role through its effect on planetary albedo.

7. Sea ice distribution has a critical feedback role and needs to be modeled correctly.

8. Permafrost and seasonal land ice have significant roles and also should be included in a comprehensive model.

Some of these points are illustrated by Manabe and Stouffer's (1980) equilibrium model results for a fourfold increase in carbon dioxide. The model gave up to 12°C cooling (at 25 km) in the stratosphere, with moderate seasonal dependency in the polar regions. The troposphere warmed by 3 to 5°C, on average, but up to 13°C at the North Pole in winter. The antarctic region troposphere, however, warmed by 6 to 7°C, with little difference between winter and summer. These results can be compared with the more recent results of Washington and Meehl (1984) using the NCAR climate model. The results are generally very comparable when allowance is made for the fact that Washington and Meehl assumed a doubling of carbon dioxide. However, the latter model predicts a much greater lower-tropospheric warming near the margin of Antarctica in winter than in summer and considerably more warming there in both seasons than that predicted by the Manabe and Stouffer model.

Returning to the difference between the stratosphere and troposphere, the troposphere and the surface are undergoing a coupled greenhouse warming process. Tropospheric carbon dioxide absorbs 15-micron radiation from the surface and re-emits at a somewhat colder temperature to higher altitudes, with the differential heat energy remaining below. To the "zeroth" order, the troposphere and surface respond as a coupled slab. The time scale of atmospheric response is primarily controlled by heat uptake in the world oceans and is of the order of decades.

In the stratosphere, cooling occurs because the loss of photons to space dominates the radiative balance, with exchange of radiation between adjacent layers of the atmosphere essentially canceling out. Vertical convective coupling plays a secondary role in the stratosphere. The carbon dioxide in the stratosphere is "optically thin" compared to carbon dioxide in the troposphere.

As to positive feedback mechanisms, the lapse rate is of particular importance in winter. Because there is strong wintertime vertical stratification of the atmosphere in high latitudes, the surface is only weakly coupled to the troposphere. Therefore, warming at the surface from greenhouse gases or any other source is largely confined to the surface. Greenhouse radiative warming is consequently concentrated near the surface. The resulting steeper thermal lapse rate means that higher atmospheric layers will radiate away less heat energy, compared to the surface, thereby allowing the surface to heat up still more until a radiative balance is achieved. In this way, steepening of the thermal lapse rate acts as a positive feedback mechanism.

The albedo of ice and snow also acts as a positive feedback. As surface temperatures rise, ice and snow cover diminish unless wintertime precipitation increases enough to maintain them. Less ice and snow cover results in less reflected solar radiation, thus amplifying the warming at the surface and quickening the ice and snow loss. The magnitude of this feedback depends on the product of surface albedo change and incident solar radiation at the surface. Therefore, seasonality plays a role, and the process is most sensitive to snow and ice that persist past the spring equinox. Since sea ice is much more persistent than land snow in spring and summer, the feedback process is mainly dependent on changes in sea ice extent.

A climate modeling experiment by Hansen et al. (1984) for doubled carbon dioxide shows a warming of 3 to 4°C in tropical latitudes at all times of the year. In contrast, the model gives a seasonal variation from 1°C warming in summer to 12°C in winter in the North Polar region and from 3°C in summer to 12°C in winter at the margins of Antarctica (70°S latitude). A similar modeling experiment by Washington and Meehl (1984) shows a generally similar dependence of warming on the season for both polar regions, but with an especially pronounced wintertime warming of up to 18°C at the margin of Antarctica. If one examines such differences in model results more closely, it turns out that most of the differences can be related to differences in the modeling of sea ice.

Manabe and Stouffer (1980) in their modeling experiment interpreted the seasonal cycle in climate warming as follows: sea ice will disappear or puddles will form in the summer. This will allow more solar radiation to be absorbed in the ocean, but the ocean surface waters will warm only slightly because of large heat capacity and some mixing. In the milder winters associated with climate warming, sea ice will form more slowly than it does under present climate conditions, and the resultant sea ice will be thinner at the same time of year. As a result, the atmospheric surface layer will be warmed above present levels, triggering the lapse rate feedback process. However, the effect will be less pronounced in late winter as the ice thickens.

Horizontal transport of atmospheric sensible and latent heat and of oceanic heat is also important. More accurate and detailed observations are needed to better determine these kinds of transport. The total heat transport is known fairly well from the radiation balance at the top of the atmosphere. Atmospheric transport estimates are probably somewhat inaccurate, and oceanic transport, computed as a residual, is even more inaccurate. It may be better to estimate ocean transports by use of surface energy balance computed from a general circulation model (GCM) that uses observed ocean temperatures.

There are two different ways to parameterize convection in the numerical models: (1) use of a moist convective adjustment process and (2) use of modeled penetrative convection. The choice of scheme can have a marked effect on the latitudinal and vertical distribution of warming due to greenhouse gases.

High-latitude cloud cover is especially important because it potentially can mask much of the ice albedo feedback process mentioned previously. Unfortunately, observations of high-latitude clouds are very sparse for a number of reasons. The GCM models probably do a poor job of simulating the actual high-latitude cloud cover because of our inability to validate the models against sufficient observations.

There has been a wide range of results for sea ice change in GCM simulations, with a consequent wide range of warming predicted for the high latitudes. It appears to be difficult to simulate present sea ice coverage correctly with GCMs, let alone have any confidence about future projections. Since sea ice coverage depends critically on ocean temperature and salinity, the latter must be accurately determined before there is any hope of obtaining the current sea ice coverage.

With respect to the role of permafrost and sea ice modeling, it has been noted that conductive heat flux into frozen ground is large enough during spring and summer to significantly cool the surface. Reduction of seasonal ice thickness and permanent ice cover should help amplify summertime greenhouse warming in the high latitudes of the Northern Hemisphere, but this hypothesis has not yet been quantitatively tested. In order to test the hypothesis, development of ground ice models and their inclusion in future GCM warming studies is needed.

In conclusion, the basic mechanisms of the high-latitude amplification of greenhouse warming are reasonably well understood. However, realistic simulations by GCMs at these latitudes are not yet available. The modeling groups that are looking at the climate change process are relatively small in size compared to the large task they have undertaken, and as a result their focus is necessarily global, so that not enough attention has been given to the special problems of the high latitudes. There is a need for the larger community that works on modeling problems to focus attention on model physics at the high latitudes.

(In response to a question): There has been little change over the last 20 years or so in the approaches of the various modeling groups that are looking at climate change. This appears to be due both to limited resources and to a tendency of the modelers to fixate on specific aspects of the total problem and not to explore other areas that need work. Although more and better observations, hopefully from satellite programs such as the earth observing system (EOS), should help, the modelers will need to be motivated to make full use of such observations.

REFERENCES

Bolin, B., B.R. Döös, J. Jager, and R.A. Warrick (eds.). 1986. How will climate change? The climate system and modelling of future climate. Chapter 5 in The Greenhouse Effect, Climatic Change, and Ecosystems. John Wiley & Sons, Chichester, SCOPE 29, pp. 207-270.

Hansen, J., A. Lacis, D. Rind, G. Russell, P. Stone, I. Fung, R. Ruedy, and J. Lerner. 1984. Climate sensitivity: Analysis of feedback mechanisms. In Climate Processes and Climate Sensitivity, J.E. Hansen and T. Takahashi (eds.), Maurice Ewing Series 5, American Geophysical Union, Washington, D.C., 368 pp.

Manabe, S., and R.J. Stouffer. 1980. Sensitivity of a global climate model to an increase of CO_2 concentration in the atmosphere. J. Geophys. Res. 85:5529-5554.

Washington, W.M., and G.A. Meehl. 1984. Seasonal cycle experiment on the climate sensitivity due to a doubling of CO_2 with an atmospheric general circulation model coupled to a simple mixed layer ocean model. J. Geophys. Res. 89:9475-9503.

Appendixes

Appendix A
Letter from the National Climate Program Office Requesting a Symposium

JAN 1 9 1988

UNITED STATES DEPARTMENT OF COMMERCE
National Oceanic and Atmospheric Administration
Rockville, Md. 20852
NATIONAL CLIMATE PROGRAM OFFICE

January 12, 1988

Dr. John Perry
Executive Secretary
Board of Atmospheric Sciences and Climate
National Academy of Sciences
2101 Constitution Avenue
Washington, DC 20418

Dear John:

As you know, the most recent observations of the Antarctic "Ozone Hole" indicate that it not only is larger in magnitude in 1987 than any previous year, but also lasting longer- presumably because of the persistence of the polar vortex which did not break up until late November (rather than late October). This situation is apparently caused by anomalous low temperatures which contribute to the formation of the polar stratospheric clouds - a necessary condition for the heterogeneous chemistry that causes the rapid ozone depletion. The ozone depletion, in turn, contributes to the persistence of the low temperatures - a positive feedback effect.

There is also the distinct possibility that the increase of methane in the atmosphere and its subsequent oxidation in the stratosphere ($CH_4 + OH \rightarrow CH_3 + H_2O$) is significantly contributing to the growth of these polar stratospheric clouds. If this is indeed the case, then the prospect of an increase in the rate of methane outgassing from Arctic tundra and permafrost due to projected greenhouse warming (enhanced in higher latitudes) will certainly exacerbate both the greenhouse warming <u>and</u> the stratospheric ozone depletion.

In view of this most serious combination of positive feedbacks, I suggest that you consider a joint planning meeting of the Climate Research and Atmospheric Chemistry Committees for the purpose of (1) assessing our current understanding of the linkages between greenhouse warming and stratospheric ozone depletion, and (2) identifying possible gaps or needs in our research programs. Now, more than ever, the need exists to provide sound scientific input to policy-makers on these issues. In this connection, the terms of the recent Montreal Agreement to limit the world wide production of CFC's provide for revising the schedule of reductions based on a re-assessment of the environmental situation.

In view of the possibility that the greenhouse effect may be accelerating the depletion of stratospheric ozone, and the probability that both are contributing to the unprecedented (see attached) decrease in Southern Hemisphere stratospheric temperature, I am requesting that this matter be given the highest priority in the BASC plan for 1988.

Sincerely,

Alan D. Hecht
Director,
National Climate Program
Office

Appendix B
Symposium Agenda and Participants

Agenda

Symposium Chairman: Richard A. Anthes

0830 Opening Remarks
—Richard A. Anthes, National Center for Atmospheric Research

0840 Introduction
—Alan Hecht, National Climate Program Office

Chairman, Morning Session: Robert Sievers, University of Colorado

0900 Global Change and the Changing Atmosphere
—William C. Clark, Harvard University

0920 Discussion

0930 Stratospheric Ozone Depletion: Global Processes
—Daniel L. Albritton, National Oceanic and Atmospheric Administration

0950 Discussion

1000 Stratospheric Ozone Depletion: Antarctic Processes
—Robert T. Watson, National Aeronautics and Space Administration

1020 Discussion
1030 Break
1045 The Role of Halocarbons in Stratospheric Ozone Depletion
—F. Sherwood Rowland, University of California, Irvine
1105 Discussion
1115 Heterogeneous Chemical Processes in Ozone Depletion
—Mario J. Molina, Jet Propulsion Laboratory, National Aeronautics and Space Administration
1135 Discussion
1145 Lunch

Chairman, Afternoon Session: Robert E. Dickinson, National Center for Atmospheric Research

1245 Free Radicals in the Earth's Atmosphere: Measurement and Interpretation
—James G. Anderson, Harvard University
1305 Discussion
1315 Theoretical Projections of Stratospheric Change due to Increasing Greenhouse Gases and Changing Ozone Concentrations
—Jerry D. Mahlman, Geophysical Fluid Dynamics Laboratory, National Oceanic and Atmospheric Administration
1335 Discussion
1345 Historical Trends in Atmospheric Methane Concentration and the Temperature Sensitivity of Methane Outgassing from Boreal and Polar Regions
—Robert C. Harriss, Langley Research Center, National Aeronautics and Space Administration
1405 Discussion
1415 Global Temperature Trends
—Kevin E. Trenberth, National Center for Atmospheric Research
1435 Discussion
1445 Break
1500 Use of Numerical Models to Project Greenhouse Gas-Induced Warming in Polar Regions (The Conceptual Basis Developed Over the Last Twenty Years)
—Robert E. Dickinson, National Center for Atmospheric Research
1520 Discussion

1530 European Research and Views
—Dieter H. Ehhalt, Federal Republic of Germany

1545 International Programs
—S. Ichtiaque Rasool, National Aeronautics and Space Administration

1600 Panel Discussion
Richard A. Anthes, *Chairman*
Eugene Bierly, National Science Foundation
Robert Dickinson, National Center for Atmospheric Research
J. Michael Hall, National Oceanic and Atmospheric Administration
Alan Hecht, National Climate Program Office
Noel Hinners, National Aeronautics and Space Administration
Robert Sievers, University of Colorado

1715 Closing Remarks, Richard A. Anthes

1745 Adjourn

1800 Reception

Participants

Members, Board on Atmospheric Sciences and Climate

R. Anthes, *Chairman*
J. Anderson
K. Bryan
R. Cicerone
A. Dessler
J. Dutton
J. Gerber
M. Glantz
T. Graedel
J. Hovermale
R. Johnson
T. N. Krishnamurti
J. Nogués-Paegle
C. Rooth
W. Washington

Members, Climate Research Committee

R. Dickinson, *Chairman*
D. J. Baker
I. Fung
A. Gordon
A. Leetmaa
J. Mahlman
E. Sarachik
J. Sarmiento
K. Trenberth
J. Walsh

Members, Committee on Atmospheric Chemistry

R. Sievers, *Chairman*
W. Chameides
D. Ehhalt
F. Fehsenfeld
R. Harriss
C. Kolb
F. S. Rowland

Guests

T. Ager, U.S. Geological Survey
D. Albritton, National Oceanic and Atmospheric Administration
J. Angell, National Oceanic and Atmospheric Administration
E. Bierly, National Science Foundation
W. C. Bolhofer, National Oceanic and Atmospheric Administration
P. Brewer, Woods Hole Oceanographic Institution
R. Cooper, Congress
T. Cremins, Institute of Security and Cooperation in Outer Space
T. Delaca, National Science Foundation
J. Diamante, National Oceanic and Atmospheric Administration
B. Doe, U.S. Geological Survey
B. Döös, National Climate Program Office, NOAA
E. Epstein, National Weather Service, NOAA
R. Etkins, National Climate Program Office, NOAA
J. Fein, National Science Foundation
W. Forster, Department of Energy
L. Gray, Jr., Chemical Manufacturers Association
L. Green, Jr., ECAL
R. Greenfield, National Science Foundation
J. M. Hall, National Oceanic and Atmospheric Administration
R. Hallgren, National Weather Service, NOAA
J. Hansen, National Aeronautics and Space Administration
L. Hanson, University of Rhode Island
A. Hecht, National Climate Program Office, NOAA
R. Hirsch, Department of Energy
J. Hoffman, Environmental Protection Agency
W. Hooke, National Oceanic and Atmospheric Administration
J. Jordan, National Science Foundation
J. Joyce, National Science Foundation
J. Justus, Congressional Research Service, Library of Congress
P. Jutro, Environmental Protection Agency
R. Keesee, National Science Foundation
J. Kermond
K. Kimball, Congress

M. Kurylo, National Aeronautics and Space Administration
J. Laurmann, consultant
R. Lavoie, National Weather Service, NOAA
J. Mahoney, NAPAP
M. Manton, Burke, Australia
R. J. McNeal, National Aeronautics and Space Administration
M. McShea, Congress
M. Molina, Jet Propulsion Laboratory
W. R. Mooman, World Resources Institute
C. Moore, Congress
J. Moyers, National Science Foundation
M. Nelson, Congress
S. Oaks, Congress
G. Ohring, National Oceanic and Atmospheric Administration
R. Pomerance, World Resources Institute
J. Rasmussen, National Weather Service, NOAA
S. I. Rasool, National Aeronautics and Space Administration
R. J. Reed, University of Washington
M. Riches, Department of Energy
N. Rosenberg, Resources for the Future
R. Schiffer, National Aeronautics and Space Administration
S. Shimberg, Congress
K. Smythe, Science and Policy Association
D. Stirling, University Corporation for Atmospheric Research
N. Strommen, U.S. Department of Agriculture
G. Tesi, National Science Foundation
R. Watson, National Aeronautics and Space Administration
G. Wetstone, Congress
H. Wiser, Environmental Protection Agency

Staff, National Academy of Sciences/National Academy of Engineering/National Research Council

K. Bergman
R. DeFries
A. Hoffman
D. Hunt
R. Kasper
M. H. Katsouros
J. Mackaness
N. Metzger
T. Milan
J. S. Perry
M. Uman
F. D. White
R. M. White

Appendix C
Glossary

Special Terms

Albedo—Reflectivity of clouds, land, ocean, ice, or snow surfaces to incident solar radiation.

Antarctic ozone hole—A substantial reduction below the naturally occurring concentration of ozone over Antarctica.

Catalytic cycle—Refers, in these proceedings, to a cycle of chemical reactions, involving several chemical compounds, that results in the destruction of ozone molecules by ionized chlorine atoms.

Climate model—A numerical simulation of the climate system. Climate models are of two basic types: (1) static, in which atmospheric motions are neglected or are represented with a simple parameterization scheme such as diffusion, and (2) dynamic, in which atmospheric motions are explicitly represented with equations. The latter category includes general circulation models (GCMs).

Dobson unit—A measure of total column atmospheric ozone. If brought to 1 atmosphere (1013.2 mb) of atmospheric pressure, 100 DU of pure ozone would measure 1 mm thick. Normal total column atmospheric ozone averages about 300 DU.

End effects—In the analysis of a time-series of observations, refers to the increased uncertainty in estimating trends near the beginning and end of a series record.

Feedback—In climate studies, the amplification (positive feedback) or dampening (negative feedback) of climate change by climatic processes that are a consequence of the change.

Greenhouse gases—Trace gases in the atmosphere that are strongly absorbent in parts of the infrared wavelength spectrum. These include carbon dioxide, methane, nitrous oxide, and some of the chlorofluorocarbons.

Heterogeneous chemistry—A category of chemical reactions that involve both gaseous and liquid ingredients.

Ice core data—Paleoclimatic temperature data deduced from chemical composition, and its variations with time, of ice samples obtained by vertical drilling in glaciers.

Lidar—Light detection and ranging system. An instrument that uses infrared or visible light in the form of a laser beam to measure wind speed and direction from the movement of wind-borne aerosols.

Mean life—Time required for concentrations to diminish to 1/e of the original value.

Microwave sounder unit—A satellite-based remote sensor capable of measuring temperature in the lower stratosphere under certain conditions.

Model physics—In climate models, the representation of physical processes, especially atmospheric radiative balance and heat exchange processes between, for example, atmosphere and ocean and atmosphere and ice.

Montreal Protocol—An international agreement to limit and eventually reduce the amount of chlorofluorocarbons injected into the atmosphere. Terms of the agreement reached in Montreal on September 16, 1987, are summarized in Chapter 3.

Ozone self-healing—Refers to the hypothesis that a reduction of ozone in the higher stratosphere would allow more ultraviolet radiation to penetrate to the lower stratosphere and create more ozone there, thus limiting total column ozone depletion.

Ozonesondes—(a) Instruments that measure the vertical profile of ozone concentration in the atmosphere; (b) vertical profiles of atmospheric ozone concentration.

Parameterization—In numerical models of the climate system, the representation of a physical process by statistical or empirical relationships rather than by equations that explicitly describe the physical process.

Polar vortex—In the stratosphere, a strong belt of winds that encircles the South Pole at mean latitudes of approximately 60°S to 70°S. A weaker and considerably more variable belt of stratospheric winds also encircles the North Pole at high latitudes during the colder months of the year.

Radiosonde—Instrument system carried aloft into the atmosphere by balloon; it measures atmospheric pressure, temperature, and humidity, and relays this information to a receiver at the launch site.

Solar maximum—The time of maximum sunspot activity during the solar cycle of approximately 11 years. The last solar maximum occurred in 1979-1980.

Solar minimum—The time of minimum sunspot activity during the solar cycle. The last solar minimum occurred in 1985-1986.

Stratosphere—Portion of the atmosphere between the tropopause (at 8- to 18-km elevation, depending on latitude and season) and the stratopause (approximately 50 km).

Umkehr network—A network of ground-based Dobson instruments that measure ozone concentrations in atmospheric layers.

Abbreviations

AER	Atmospheric and Environmental Research, Inc.
CFC	Chlorofluorocarbon. The primary CFCs affecting ozone concentrations are CFC-11 (trichlorofluoromethane) and CFC-12 (dichlorodifluoromethane).
CFM	Chlorofluoromethane-type compound. See CFC.
CGC	Committee on Global Change.
DU	Dobson unit (see special terms).
ECMWF	European Centre for Medium-Range Weather Forecasts
EOS	Earth observing system.
GAGE	Global Atmospheric Gas Experiment.
GCM	General circulation model (see climate model).
GFDL	Geophysical Fluid Dynamics Laboratory, NOAA.
IGBP	International Geosphere-Biosphere Program.
IGY	International Geophysical Year.
JPL	Jet Propulsion Laboratory.
mb	Millibar (of atmospheric pressure). One mb equals 1,000 dyne/cm^2 or 100 Pa.

NASA	National Aeronautics and Space Administration.
NCAR	National Center for Atmospheric Research.
NMC	National Meteorological Center, NOAA.
NOAA	National Oceanic and Atmospheric Administration.
NSF	National Science Foundation.
PSCs	Polar stratospheric clouds. These ice-crystalline clouds have been observed in polar regions in winter and spring.
QBO	Quasi-biennial oscillation. An approximately 26-month periodic reversal of equatorial stratospheric winds between easterly and westerly direction.
SAGE	Stratospheric Aerosol and Gas Experiment (SAGE I and II). Refers to experimental remote-sensing instruments aboard satellite platforms.
SBUV	Solar backscatter ultraviolet (satellite instrument).
TOMS	Total Ozone Mapping Spectrometer. A remote sensor mounted on the NIMBUS-7 satellite that measures total column ozone.
TOVS	TIROS operational vertical sounder.
UV	Ultraviolet (wavelength of electromagnetic radiation).
WMO	World Meteorological Organization.

Index

E

F

G

H

I

K

L

T

U

V

W